The Palgrave Macmillan Animal Ethics Series

Series Editors
Andrew Linzey, Oxford Centre for Animal Ethics, Oxford, UK
Clair Linzey, Oxford Centre for Animal Ethics, Oxford, UK

In recent years, there has been a growing interest in the ethics of our treatment of animals. Philosophers have led the way, and now a range of other scholars have followed from historians to social scientists. From being a marginal issue, animals have become an emerging issue in ethics and in multidisciplinary inquiry. This series will explore the challenges that Animal Ethics poses, both conceptually and practically, to traditional understandings of human-animal relations. Specifically, the Series will:

- provide a range of key introductory and advanced texts that map out ethical positions on animals;
- publish pioneering work written by new, as well as accomplished, scholars;
- produce texts from a variety of disciplines that are multidisciplinary in character or have multidisciplinary relevance.

For further information or to submit a proposal for consideration, please contact Amy Invernizzi, amy.invernizzi@palgrave-usa.com.

Philip J. Sampson

Nonconformist Perspectives on Animals and Language

Speaking for the Voiceless

palgrave
macmillan

Philip J. Sampson
Oxford Centre for Animal Ethics
Oxford, UK

ISSN 2634-6672 ISSN 2634-6680 (electronic)
The Palgrave Macmillan Animal Ethics Series
ISBN 978-3-032-12157-8 ISBN 978-3-032-12158-5 (eBook)
https://doi.org/10.1007/978-3-032-12158-5

This Palgrave Macmillan imprint is published by the registered company Springer Nature Switzerland AG
The registered company address is: Gewerbestrasse 11, 6330 Cham, Switzerland

"Speak up for those who cannot speak. Speak for the rights of all those who are defenceless." Prov. 31.8

"A beast cannot speak to move us to pitie and compassion; and therefore we must go to him of our own good will, though we be not moved or requested there-unto". John Calvin (1583)

I speak on "behalf of a race of poor friendless beings who certainly cannot speak for themselves". Sir Richard Hill, opposing bull baiting (1802).

Series Editor's Preface

This is a new book series for a new field of inquiry: Animal Ethics.

In recent years, there has been a growing interest in the ethics of our treatment of animals. Philosophers have led the way, and now a range of other scholars have followed from historians to social scientists. From being a marginal issue, animals have become an emerging issue in ethics and in multidisciplinary inquiry.

In addition, a rethink of the status of animals has been fuelled by a range of scientific investigations which have revealed the complexity of animal sentiency, cognition, and awareness. The ethical implications of this new knowledge have yet to be properly evaluated, but it is becoming clear that the old view that animals are mere things, tools, machines, or commodities cannot be sustained ethically.

But it is not only philosophy and science that are putting animals on the agenda. Increasingly, in Europe and the United States, animals are becoming a political issue as political parties vie for the "green" and "animal" vote. In turn, political scientists are beginning to look again at the history of political thought in relation to animals, and historians are beginning to revisit the political history of animal protection.

As animals grow as an issue of importance, so there have been more collaborative academic ventures leading to conference volumes, special journal issues, indeed new academic animal journals as well. Moreover, we have witnessed the growth of academic courses, as well as university posts, in Animal Ethics, Animal Welfare, Animal Rights, Animal Law, Animals and Philosophy, Human-Animal Studies, Critical Animal Studies, Animals and Society, Animals in Literature, Animals and Religion—tangible signs that a new academic discipline is emerging.

"Animal Ethics" is the new term for the academic exploration of the moral status of the non-human—an exploration that explicitly involves a focus on what we owe animals morally, and which also helps us to understand the influences—social, legal, cultural, religious, and political—that legitimate animal abuse. This series explores the challenges that Animal Ethics poses, both conceptually and practically, to traditional understandings of human-animal relations.

The series is needed for three reasons: (i) to provide the texts that will service the new university courses on animals; (ii) to support the increasing number of students studying and academics researching in animal related fields, and (iii) because there is currently no book series that is a focus for multidisciplinary research in the field.

Specifically, the series will

- provide a range of key introductory and advanced texts that map out ethical positions on animals;
- publish pioneering work written by new, as well as accomplished, scholars, and
- produce texts from a variety of disciplines that are multidisciplinary in character or have multidisciplinary relevance.

The new Palgrave Macmillan Series on Animal Ethics is the result of a unique partnership between Palgrave Macmillan and the Ferrater Mora Oxford Centre for Animal Ethics. The series is an integral part of the mission of the Centre to put animals on the intellectual agenda by facilitating academic research and publication. The series is also a natural complement to one of the Centre's other major projects, the *Journal of Animal Ethics*. The Centre is an independent "think tank" for the advancement of progressive thought about animals, and is the first Centre of its kind in the world. It aims to demonstrate rigorous intellectual enquiry and the highest standards of scholarship. It strives to be a world-class centre of academic excellence in its field.

We invite academics to visit the Centre's website http://www.oxfordanimalethics.com and to contact us with new book proposals for the series.

General Editors

Andrew Linzey
Clair Linzey

Preface

> Speak out on behalf of the voiceless, and for the rights of all who are vulnerable. Prov. 31.8-9

I have argued elsewhere that an identifiable body of animal friendly discourse emerged in England from the sixteenth century; I termed this "nonconformist". Always subordinate, it nevertheless resisted the dominant way of talking about animals, and had a disproportionate influence on pioneering law reform and the founding of welfare organisations such as the RSPCA.

This book expands that narrative, and looks in more detail at the way that the nonconformist language of resistance said novel things about the place of animals in specific human activities and institutions. It explores the nonconformist vision of creation as praise, and the idea that animals, with humans, participate in their own distinctive ways in a song of praise to God. So Isaac Watts, the nonconformist hymn writer, exhorts "let heaven and nature sing". Indeed, according to the nonconformists, it was this participation in praise that gave meaning to the lives of animals, and this also gave rise to a distinctive understanding of animal cruelty. I argue that this provides resources to look anew at the way humans treat animals.

An understanding of creation as song provides the structure for the chapters that follow, as we explore the idea of the harmony of creation, the calling of human beings to "conduct" creation's song, and ways in which cruelty to animals, in contexts such as hunting and warfare, prevents them from fulfilling their calling to praise God.

The book concludes by exploring how, during the twentieth century, a new evangelicalism abandoned this distinctive perspective inherited from their forebears, and adopted a "carno-theology" to justify their participation in the cruel practices associated with industrial animal farming and slaughter.

Animal Friendly Pioneers

In his seminal study of 1983, *Man and the Natural World*, Keith Thomas (1983, 153, 180) noted the emergence of a more friendly way of talking about animals in sixteenth-century England. It was pioneered by Puritans, Dissenters, Quakers, and Evangelicals; the bible, not classical authors, was the "authority most frequently cited". A prominent feature, remarkable for its time, was its opposition to animal cruelty. This list of pioneers is not exhaustive. We must certainly add eighteenth-century Methodism. John Lawrence (1796, vol. 1, 120) wrote in the 1790s of his concern that he would be thought a Methodist for his kindness to animals. Calvinistic Baptists of the nineteenth century such as C. H. Spurgeon were also outspoken in what we would now call "animal advocacy". Retrospectively, we should also include the European Reformers whose texts were seminal in English Protestantism. Not only did this discourse influence law reform and the founding of welfare societies, but vestiges persist even in today's way of talking about animals. I have previously referred to this body of literature, and literary culture, as "nonconformist" (Sampson 2018, vii–x).

In the half century since Thomas' pioneering work, many authors have noted what Bruce Boehrer (2007, 26) calls a "sympathetic identification with the suffering of animals" from the sixteenth century. Moreover, this was not just a historical curiosity, but has had an enduring impact. Rod Preece and Lorna Chamberlain (1993, 34) attribute the success of animal welfare legislation in Britain, not to Jeremy Bentham or philosophers, but as stemming "directly from the humanitarian impulses which lay behind the evangelical Protestantism of the period". Belden Lane (2011, 33; also 32, 223) notes the "significant Puritan influence on the modern development of animal rights".

Despite this growing recognition, Margaret Adam (2019, 7) still considers that "most descriptions of the passing of legislation against cruelty towards animals in the early to mid-nineteenth century exaggerate the contributions of Jeremy Bentham's utilitarianism and ignore the roles of Christian arguments and evangelical Christians in Parliament". Indeed, it is still commonplace to hear that the human exploitation of animals, causing routine and terrible suffering, is a consequence of biblical doctrines of ***dominion*** and "our Divine Creation in the ***image of God***", both allegedly giving us absolute authority to use animals as we wish (Singer 2011, 64—my emphasis). Yet *dominion* and the*imago dei* are central to the nonconformist literature identified as pioneering a more animal friendly discourse. This seeming paradox is addressed in Chapter 1.

People of the Book

Nonconformists, in the sense described above, were a diverse group. Moreover, they were always a marginal minority, sometimes expelled from the established church and unable to participate in university life or politics. At other times they were central to the events of their day. Yet even at their most influential, they were regarded with suspicion, and subjected to ridicule by the dominant discourse. There is no single articulation of nonconformist discourse to which to point, but an overlapping family of meanings which has repeatedly been identified in the more animal friendly ways of speaking.

Despite their many differences, their texts show a remarkable consistency, what Thomas (1983, 153) called "little historical development", in their way of talking about animals. In particular, this arose from a common textual source. They were "people of the book", and that book was the bible; their writings were "thickened with scripture" as the puritan John Trapp somewhere said of their prayers. Nonconformists commonly read the bible daily, and recognised in it their own experience of animal husbandry. They found there multiple points of resistance to the then dominant ways of talking about animals (Sampson 2018 ch. 5–7). It is nowadays difficult to appreciate the significance of this to a general public which was, by 1600, the most literate in Europe; most having access to a bible. Many biblical narratives and themes "were absorbed by the popular mind, unoccupied for the most part by any rival learning". Henry VIII complained to Parliament that the bible was being "disputed, rhymed, sung and jangled in every ale house and tavern" in England (Bobrick 2001, 275–77. Hall 2019, 14f). This applied to animal friendly ways of speaking as much as to anything else.

Moreover, nonconformists did not live in ivory towers. They influenced public opinion through sermons, Sunday schools, and popular books, from John Bunyon to William Copwer. Sometimes deeply involved in the events of their times, they included some of the foremost scholars of the day, and many were responsible for local congregations. Nonconformity, and especially Puritanism, has today acquired a reputation for dour austerity, making it hard to understand how they could have influenced anybody. This is entirely mistaken. For example, Richard Baxter (1817:1650, 96) declared that he is the "best Christian... who has the readiest passages, not so much from the ear to the brain, as from that to the heart". "Take no nay of him", advised Richard Sibbes (1862:1638, vol. 2, 206), speaking of Christ, "till he hath given thee the kisses of his love". The influence of puritans such as these stirred the soul, especially compassionate souls.

A Practical Piety

Of course, nonconformity shared much with its contemporaries, both about their world in general, as well as specific social and political affiliations. This shared context included the then prevailing indifference to animal cruelty. That many individual nonconformists adopted stances widely held by the general public at the time need not surprise us. This book will be concerned, not with similarities, but with the differences, with views which had little support among their contemporaries. In particular, with nonconformist resistance to, and even subversion of, the dominant discourse of their day in relation to animals. Their language spoke of the mercy, kindness, and compassion which are owed to animals who have a voice to praise their creator, and a dignity to manifest Him in the world.

This novel language arose partly from the break with medieval "spirituality" which was linked to monastic living. Nonconformists tended to speak of "piety" rather than "spirituality", of existential integrity of life rather than mystical withdrawal from it. When they read of animals in their bibles, they thought of the real animals who shared the lives, sometimes the homes and food, of their congregations. They knew that it was not possible to over-winter all the animals born in spring, and they faced the same practical issues as the farmers in their bibles. Their reading of the bible was rooted in this knowledge. There are many biblical pericopes which describe the lives and deaths of animals. In order to bear the theological weight they carry in both Judaism and Christianity, they must first make sense within the animal husbandry of their day (Douglas 1999). This does not diminish the theology they bear; quite the opposite. To recognise the life-giving and cleansing role of water in the everyday life of the Ancient Near East does not disqualify its significance in baptism; rather, it makes possible and enriches the theological meaning of baptism. So also with animals.

Saying New Things

This novel language of animals enabled nonconformists to say new things about human-animal relationships. Cruelty was part of early modern life. Nonconformist farmers knew it is wrong as a matter of experience and conscience. They also knew that it is an offence against God; the psalmist calls upon animals to praise their creator, not on men to make them scream (Ps 148, 150 etc). Nonconformist piety understood their everyday experience of animals clothed in these biblical narratives. This book explores the consequences of that.

The opening chapters set out the nonconformist vision of creation as manifesting and praising God. Thomas Ken's well known doxology begins "Praise God from whom all blessings flow, praise Him *all creatureshere below*". Many Christians regularly sing it still. In announces our human calling to conduct animals in their song of life, not consume them in a carno-theology of death. For nonconformists, this was not simply a "spiritual" thought to accompany singing the doxology in church, but a practical demand upon how they lived a life of piety. It had implications for specific human activities and institutions: for blood sports such as hunting; for warfare in which animals are "burnt in the stall"; and for meat-eating which attracts abuse and excess (Abbot 1613, 453). The second half of this book explores those implications. I argue that nonconformist discourse provides resources to look at the way humans treat animals from a perspective now foreign to most of us, and to see them anew.

Texts and Contexts

Many of the texts cited in this book date from a time before the standardisation of spelling and grammar. I have modernised typography, but have otherwise not altered the English texts cited save to provide an explanation of archaisms. Some texts have lengthy titles and subtitles. I have generally shortened these or omitted excessive elaboration, using ellipsis to indicate a contraction. Many nonconformists read the biblical texts in the original languages, but where they employed a translation it was almost always that of 1611, now known as the Authorised Version. Where I quote the bible, it is in this translation which I abbreviate to "the AV". I generally employ capitalised pronouns when referring to deity, and to indicate that words such as "Fall" are to be understood in their theological sense.

Finally, I distinguish between *evangelical* and *evangelicalism*. *Evangelical* refers principally to the eighteenth-century Evangelical Revival and its nineteenth-century echoes (Lloyd-Jones 1992, 12f). *Evangelicalism* refers specifically to the emergence from the later nineteenth century of an individualistic and anthropocentric departure from the nonconformist tradition. Stewart (2005, 152) dates the use of *evangelicalism* from 1831 in a similar sense to that used here. Bebbington (1989, 20) argues for an earlier transformation of Reformation and Puritan discourse into evangelicalism, dating from the mid-eighteenth century; a more phenomenological than discursive emphasis. In Chapter 9, I argue for a gradual transition, marked by C. H. Spurgeon's protest in 1887 at the "downgrading" of "evangelical orthodoxy".

For the sake of brevity, I shall refer to non-human animals as "animals" or, in an early modern context, as "beasts". I have also used gendered language when referring to sources. There are disadvantages to these decisions, but the alternatives seem to me to bring more troubles than they relieve. In tracing the nonconformist discourse of animals, I often use their own language and phraseology; this should be understood as the voice of my sources, not my own. It may be thought that my use of personal pronouns for animals is an exception to this general approach, a concession to modern usage. However, it has nonconformist precedent, not least in the AV. For example, both the puritan Joseph Caryl (1670, vol. 4, 209, 212) and the magisterial reformer John Calvin (1583, 755), write "the oxe knoweth *his* owner, and the asse*his* master's crib…" (Is.

1.3), as does the AV, the first in the original English, and the second in the English translation by Arthur Golding of 1583.

Over the years, many friends and colleagues have contributed to my thinking on this subject, either as collaborator or by sharpening my thought on the stone of challenge; I am grateful to them all. In particular, I would mention the participants at the Oxford Centre for Animal Ethics Summer-schools. My greatest debt is to my wife, Dr. Miriam Sampson whose support and sociological imagination have been invaluable. Finally, I express my sincere gratitude to Prof. Andrew Linzey for his constant encouragement. Naturally, all shortcomings are my own.

Oxford, UK Philip J. Sampson

References

Abbot, George. 1613. *An Exposition upon the Prophet Jonah*. London: Richard Field.

Adam, Margaret B., David L. Clough, and David Grumett. 2019. A Christian Case for Farmed Animal Welfare. *Animals* 9: 1–13.

Baxter, Richard. 1817:1650. *The Saints' Everlasting Rest* Glasgow: Blackie.

Bebbington, D. W. 1989. *Evangelicalism in Modern Britain: A History from the 1730s to the 1980s*. Abingdon: Routledge.

Bobrick, Benson. 2001. *The Making of the English Bible*. London: Weidenfed and Nicolson.

Boehrer, Bruce. 2007. *Cultural History Animals in the Renaissance*. Oxford: Berg.

Calvin, John. 1583. *Sermons on Deuteronomy*. Trans. Arthur Golding. London: John Harrison.

Caryl, Joseph. 1670. *An Exposition with Practical Observations upon the Book of Job*, vol. 4. London: Thomas Sawbridge.

Douglas, Mary. 1999. *Leviticus as Literature*. Oxford: Oxford University Press.

Hall, David D. 2019. *The Puritans*. Princeton University Press.

Lane, Belden C. 2011. *Ravished by Beauty: The Surprising Legacy of Reformed Spirituality*. Oxford: Oxford University Press.

Lawrence, John. 1796. *A Philosophical and Practical Treatise on Horses and on the Moral Duties of Man towards the Brute Creation*. London: Longman.

Lloyd-Jones, D. M. 1992. *What is an Evangelical?* London: Banner of Truth.

Preece, Rod, and Lorna Chamberlain. 1993. *Animal Welfare and Human Values*. Ontario: Wilfrid Laurier University Press.

Sampson, Philip J. 2018. *Animal Ethics and the Nonconformist Conscience*. Palgrave Macmillan.
Sibbes, Richard. 1862:1638. The Spouse, Her Earnest Desire after Christ her Husband. In *The Complete Works*, vol. 2, 198–208. Edinburgh: James Nichol.
Singer, Peter. 2011. *Practical Ethics*. Cambridge: Cambridge University Press.
Stewart, Kenneth J. 2005. Did Evangelicalism Predate the Eighteenth Century? *Evangelical Quarterly* 77 (2): 152–1533.
Thomas, Keith. 1983. *Man and the Natural World*. London: Allen Lane.

Contents

CHAPTER 1

Introduction

Wonder and Joy

This book is about speaking for animals. In particular, it aims to express their longing to manifest wonder and joy, and our office as humans to facilitate it.[1]

We do not usually talk of speaking *for* animals in this way; more commonly we speak *of*, them, or *about*, them. We do this in a variety of ways, some of which I will discuss in this book. The dominant way of talking uses the language of property; whether that of pet *ownership*, or of commercial live-*stock*. Subordinate discourses include: talking *about* them as part of the natural world, "wild", rather than owned; or speaking of them as sentient beings with capacities (such as planning

[1] The word "manifest" appears often in this book. It is important, then, to distinguish it from the current twenty-first century sense of making something a reality in your life by believing and acting as though it were already so. Rather, I shall use the word in its seventeenth century sense to mean the sensible expression, accommodated to the human understanding, of something not otherwise evident; specifically, the character or attributes of God. In this sense, God manifests himself in creation, publicly, truly and justly, though not exhaustively. Creation is, as the American congregationalist, Jonathan Edwards (1974:1765, 119), put it: "the emanation and true external expression of God's internal glory and fulness" to "created understandings". It is the existential precondition for George Herbert's (1633, 178-9) plea: "Teach me, my God and King, in all things thee to see…". "Office" is used in the sense of a position of authority and responsibility.

P. J. Sampson, *Nonconformist Perspectives on Animals and Language*, The Palgrave Macmillan Animal Ethics Series,
https://doi.org/10.1007/978-3-032-12158-5_1

ahead) and skills (such as tool making) formerly attributed only to humans (Evans et al., 2024; Stone 2018, 148–50). Only occasionally does our language entail speaking *for* them. As individuals we might do this when we seek the interests of our companion animal, as of any other family member. But the best known way of speaking *for* animals is the advocacy of various welfare and animal rights organisations. Many authors have noted that this distinctively animal friendly discourse emerged from the sixteenth century in Northern Europe, especially in England, and I have argued elsewhere that it originated in "nonconformist" Christian discourse (Sampson 2018). In this book I will draw extensively upon the nonconformist tradition to develop a way of speaking *for* animals; and apply it to a range of practical issues.

Of course, the dominant language of animals as property is not the only way of denying them a voice. There are also ways of speaking *for* animals without giving them a voice; ways which give them a voice which is not their own; and ways of speaking *for* them so that they will not speak *for* themselves. This is a risk whenever the more powerful or included speak for those with less privilege. William Cowper (1849, 599) describes a bullfinch who is taught to sing in a voice given to him by his owner:

> [...] though by nature mute
> Or only with a whistle blest,
> Well-taught, he all the sounds express'd
> Of flagellet or flute.

More violent examples include fighting animals said to enjoy the contest, bulls that are said to enjoy being baited, and horses who supposedly enjoy warfare (Hemingway 1923, 106).[2]

Only when we recognise that their voice is not being heard, that advocacy is required, can we authentically speak *for* other animals. We are more likely to recognise this if we listen, and enter into conversation with them (Jorgensen 2025, s.6). Reality is not silent, nor does it yield truth only by force; rather it reaches out to us and seeks our care (Polanyi 1974). It is full, and rich with meaning.

As for "wonder" and "joy", some ways of speaking about animals are more conducive to using such words than others. The language of ownership or property invokes little "wonder", although there may be a kind

[2] See Chapters 5-7 below.

of perverse joy in amassing great wealth, perhaps including animals or animal trophies.

Wonder

"Wonder" is more common when speaking of "wild" animals than of the domesticated cow or even the family dog. Indeed, wonder is the mainstay of many Nature documentaries, from Chris Packham's *The Wonder of Animals* to David Attenborough's *Blue Planet*. We marvel at migratory birds (how *do* they find their way?); at the soaring eagle and colour-shifting chameleon; at the metamorphosis of caterpillar into butterfly; and at the lithe speed of a cheetah, the leaping of a dolphin, or the roaring of a lion. Wonder takes us beyond the merely factual being of these animals. It also takes us beyond ourselves. The very nature of wonder entails disclosing or making manifest something which somehow transcends both animals and ourselves as autonomous beings; it opens a wider, richer, more beautiful, and meaningful world to us. We cannot generate wonder; it can be received only as a gift, and it gives an experience immeasurably greater than we can anticipate.

Joy

The experience of "joy" similarly transcends us (Lewis 1955). It seems quintessentially a human experience, yet it differs from mere pleasure or happiness in pointing beyond our humanity. We look at Nature, as Wordsworth (1798) puts it, "with an eye made quiet by the power of harmony, and the deep power of joy, we see into the life of things". Even the joy of sharing our lives with a beloved companion animal reaches beyond us as individuals to embrace our companion. Otherwise it would be mere company, like that of a familiar radio programme.

Unlike the language of nature documentaries or that of companion animals, some ways of talking about animals require the suppression of wonder and the perversion of joy, reducing the former to solely material being, and the latter to mere pleasure. The language of ownership already entails this; this is at its clearest in the owner's total power over the body of the owned. When Dostoevsky (1900, 285, 6) recounts the lashing of an over-burdened horse on her "weeping, on [her] meek eyes", it is clear that the owner did not marvel in wonder at her fortitude and courage in trying to pull her burden. It is easier, though disagreeable, to imagine that her abuser obtained a perverse joy from his "intoxication of cruelty". Indeed, the suppression of wonder and the perversion of joy is common in the language of cruelty towards animals more generally. It is also found in discourses of carnism[3]. The carnist's perverse joy at sinking his teeth into a succulent steak has entirely suppressed the wonder that a living bull might inspire. Only the material being of flesh, and epicurean pleasure, remain. In consequence, the contemporary language of both cruelty and carnism is usually anthropocentric, concerned with our human "intoxications" and pleasures. Only rarely are there echoes of an earlier age, as when we describe cruelty as evil because a less transcendent language seems inadequate to the case. What we can say depends upon how we speak.

Speaking About Animals

The way we speak about animals "is never just words", but always relies upon, and discloses our material relations with them (Žižek 2023). Sometimes our ways of speaking about animals interact in complex ways, conflicting with or displacing one-another. Some words may be specific to a particular way of speaking which their use both indexes and conjures. An example is the distinction between a "cow" and "beef". Saxon farmers bred and cared for *cows;* Norman French conquerors ate them, with *beef* ("boeuf") on the menu. Words can carrying a history of such oppression, indexing power relationships over both human bodies (Saxon farmers) and animal bodies (here, cows; but also sheep, pigs, and chickens). For the Saxon farmers, often living in conditions of scarcity, their cows were individuals known to them, usually too valuable to slaughter. For the Norman conquerors, the whole point of *beef* animals is to slaughter and eat them. To this day, such Norman derived words ("beef", "mutton", "pork", "capon") index animals as meat, as interchangeable bodies and cuts. Language has marked the transformation of living, breathing, mooing creatures into an un-nameable, shrink wrapped package of flesh. For this reason, it is perilous to give a personal name to an animal raised for the table. Rhoda Wilkie recounts the story of a farmer who abandoned her meal on being reminded that she was eating the leg of a pig which

[3] I reserve the term "carnism" for the industrial pattern of animal rearing and consumption which emerged in the late nineteenth century; see Chapter 8.

her family had named Ophelia; at that moment, gastronomic language was transformed into that of a familiar life whose dismembered body was suddenly, and sickeningly, grievable (Wilkie 2010, 155–6). Words can do things (Austin 1975). It is these conflicts and transformations within our ways of speaking about animals that allow us to grieve for the loss of a companion animal, wonder at a blue whale, and eat a beef steak.

All native speakers know when it is appropriate to use one language rather than another. Everyone knows that it is polite to remark on their host's tasty hog-roast, but not to ask how their dog Rover would taste. Western outrage at the Yulin Dog Meat Festival is fed by a failure to keep just these different discourses apart. The reclassification by China's Ministry of Agriculture of dogs as "pets" rather than "livestock" may mark the Festival's demise (Anon 2020).

Speaking for the Voiceless

Reality as Praise

So far, I have rather informally referred to different ways of speaking about animals. For example, I have mentioned dominant and subordinate discourses, the conflicts between them, and the displacement of one by another. These issues are discussed in detail in Chapter 1 which sets out a framework for exploring what can be said about animals. The chapter also introduces the "nonconformist" literature on which I extensively draw. In that theocentric discourse, the voice of all animals, and the meaning of their lives, is found in their ability to praise God, not in supposed similarities with human capacities or qualities.

In his seminal work *Man and Nature*, the historian, Keith Thomas (1983, 150f), was the first to clearly identify the more animal friendly way of speaking which emerged from the late sixteenth century in England, and to link it with a nonconformist literary culture. I have elsewhere analysed the structure and content of nonconformist discourse in relation to animals (Sampson 2018). Chapter 2 of the present book takes up that narrative, and explores its implications for the significance of animal voices. In nonconformist discourse, animals are no longer envisaged as possessing an autonomous being, but as dependent creatures, longing to manifest wonder and joy. Their voice points beyond themselves, and responds to their creator's call to praise with "song" rather than speech, expressing rather than representing. Creation discloses its creator as music discloses the heart of its composer.

Chapter 2 treats songs of praise by individual animals, but employs various images, such as those of a choir or an orchestra, which suggest a harmony of voices singing together. Chapter 3 takes up these themes, tracing a shift from the *being* of living creatures to their *meaning*. Rather than existing for the sake of humans in the hierarchy of a Great Chain of Being, animals find their meaning in harmoniously singing together. The chapter focusses on the mutual dependence of creatures within *covenants*. This opens a space to discuss both the value of individual animals and the communion of all creatures, foreshadowing the "modern idea of the balance of nature" (Thomas 1983, 278). Creation "singing together", weaving diversity into harmony, gifted the nonconformists a point of resistance to the dominant and exploitative discourses of property and instrumental use.

The nonconformist portrayal of animal reality as song raises obvious questions about the place and office of humans. The dominant discourse is anthropocentric, regarding animals as human property. Subordinate discourses place humans as one animal among others. Chapter 4 invites the reader to consider the question of being authentically human within a nonconformist discourse. Rather than ruling as head of a hierarchy, or being one animal among many, nonconformists envisaged humans as fellow musicians, conducting creation's song; our office is to facilitate the longing of other animals to manifest wonder and joy. This opened a space in which a novel ethic towards animals could consolidate. The chapter concludes by noting that cruelty cuts to the marrow of such a vision by destroying wonder and joy, save in an entirely perverse sense.

So far, chapters have discussed the nonconformist understanding of non-human animals and their relationships with humans. I have mentioned some practical consequences in passing, but have not discussed them in detail. Chapter 5 is transitional in both setting out the nonconformist approach to cruelty, and in providing some practical examples of nonconformist advocacy. From Chapter 6 on, the focus turns to specific examples of animal-human interaction.

Resisting Cruelty

Nonconformists were passionate about creation praising God, yet they were not blind to the cruelty they saw all around them. They heard the dissonances, even as they insisted on the harmony. For them, it is humans who have marred the harmony of creation; it groans, but it also looks towards restoration (Rom. 8.20f). How, they asked, do humans conduct a groaning choir in the knowledge that something is deeply amiss, and that we ourselves have caused the pain. That question is addressed in Chapter 5, which opens by contrasting the idea that suffering is normal in Nature, with the radical nonconformist view that it is an entirely abnormal consequence of human activity. Moreover, since we caused pain in the first place, it is egregious wickedness to collude in the cacophony of cruelty; even worse to add to the suffering for "sport" or convenience. Tone deaf humans, brought up to cruelty, must learn kindness. This view has proved prophetically apt in our world dominated by human cruelty. The chapter concludes with a discussion of gendered narratives of violence.

Advocating for Animals

Chapters have so far surveyed the nature of reality as praise; the harmony of singing together; the unique role of humans as conductors; and the practice of cruelty. The remainder of this book fleshes out specific examples, drawing upon the nonconformist literature.

Many readers will be familiar with the use of religious language in reforms in nineteenth-century England. William Wilberforce used it to advocate for the abolition of enslavement, but also in Parliamentary debates on the protection of animals from cruelty. Richard Hill is less well remembered, but was known in his day for his appeals to biblical language in opposing bull baiting. Similarly, social reformer, Lord Shaftesbury, made radical use of religious language in his opposition to vivisection. The remaining chapters of this book will look in detail at the way that specifically nonconformist religious language was used. Each chapter will focus on a contemporary practice of the twenty-first century which is associated with the most extreme forms of animal abuse: hunting; warfare; and eating animal flesh. I begin with a discussion of sport hunting.

Most animal advocates consider animal predation to be normal in the sense that it is a lamentable feature of "nature", and an inevitable consequence of the evolutionary struggle for survival. However, although humans are also part of Darwinian nature, sport hunting is considered ethically indefensible rather than lamentably natural. Chapter 6 explores the consequences of understanding creation as harmonious song for both predation and hunting. It conceives of normative human behaviour as modelled upon shepherd dominion, not the violent domination of predation or hunting. In particular, nonconformist discourse had little good to say about bringing up children, usually boys, to the gendered discourse of fighting, masterful virtues, as exemplified in "sport" hunting. The chapter concludes with a discussion of the appropriation of nonconformist discourse by modern anthropocentric "evangelicalism".

Hunting's development of fighting, masterful virtues made it an ideal training ground for military service. In particular, horses and the skills of horsemanship were essential for warfare before the twentieth century. Nonconformists therefore found it notable that biblical texts forbade the king of ancient Israel ownership of many horses, denying to ancient Israel the most effective weapon of their age. Chapter 7 takes up the connections between hunting and warfare, leading to an exploration of the use of animals in warfare more generally. It traces first, the nonconformist resistance to the use of animals in producing the instruments of war; and then, its resistance to the reproduction of a warrior spirit. The harmony of creation inspired a way of singing which was radically different from that of the discordant "military-animal industrial complex", founded in the cries of animals. Nonconformist discourse anticipates a future when nations learn peace, not war. Men shall "beat their swords into ploughshares", not butcher's knives (Is. 2.4). Isaiah's image suggests that the end of the suffering of animals in the military-animal industrial complex coincides with the end of their suffering in the industrial-meat complex. The next two chapters follow that lead.

Hunting and warfare are associated with extreme suffering, but both are rivalled if not exceeded by the practices of cruelty and death which constitute the modern industrial-meat complex. Chapters 8 and 9 set out, respectively: the nonconformist framework for eating animals in a subsistence culture; and the collapse of that framework into "carno-theology" in the industrialised society of the nineteenth century.

In nonconformist discourse, animals are to manifest and praise God; but subsistence farming required slaughter and meat-eating for survival. How, if we eat the choir, can we also conduct the song? Chapter 8 sets out the nonconformist discourse which addressed this question, noting that it subverts the dominant discourse of flesh-eating at a number of crucial points. In particular, animals were affirmed as created to praise God, cruelty towards them was condemned, and the "pleasures" of meat-eating were tainted by sin.

By the mid-nineteenth century, patterns of flesh-eating in England were changing as a result of urbanisation and the industrialisation of meat production, moving farther and farther away from the biblical narrative. Some nonconformists attempted to address this, but were ultimately unable to meet the challenges faced. Chapter 9 traces the collapse of the nonconformist discourse of meat-eating, and its transformation into a novel way of speaking using "proof-texts" drawn from the bible. I term the discourses surrounding these proof-texts and their exegesis in the context of diet "carno-theology". I examine its use in two communities: among Christians, especially the new anthropocentric "evangelicalism", to justify a practice which might otherwise seem deeply problematic; and its secular use to oppose meat-eating by associating it with an "obviously flawed" religious ideology.

The book concludes with a chapter recapitulating the main themes and proposing a path of witness, care, and hope.

References

Anon. 2020. China's Dogs to be Recognised as Pets Not Livestock Under Coronavirus Regulation Review. *ABC News*. https://www.abc.net.au/news/2020-04-10/china-reclassifies-dogs-as-pets-not-livestock/12140312?utm_campaign=abc_news_web&utm_content=link&utm_medium=content_shared&utm_source=abc_news_web. Accessed 11.9.2025.

Attenborough, David. 2001. *The Blue Planet*. BBC TV.

Austin, J. L. 1975. *How To Do Things With Words*. OUP.

Cowper, William. 1849. On the Death of Mrs Throckmorton's Bullfinch. In *The Works of William Cowper*, ed. T. S. Grimshaw. London: William Tegg.

Dostoyevsky, Fyodor. 1900. *The Brothers Karamazov*. New York: The Modern Library.

Edwards, Jonathan. 1974:1765. Dissertation On The End For Which God Created The World. In *The Works of Jonathan Edwards*, ed. Edward Hickman. Edinburgh: The Banner of Truth.

Evans, Louise, et al. 2024. Whoa, No-Go: Evidence Consistent With Model-Based Strategy Use in Horses During An Inhibitory Task. *Applied Animal Behaviour Science*. 277 August 2024.

Hemingway, Ernest. 1923. *Death in the Afternoon*. London: Jonathan Cape.

Herbert, George. 1633. The Elixer. In *The Temple*. Cambridge: Thomas Buck.

Jorgensen, Anders Skou. 2025. Care in Times of Crisis: Phenomenological, Political and Theological Perspectives. *Dialogue* 64 (1): 36–44.

Lewis, C. S. 1955. *Surprised by Joy*. NY: Harcourt Brace.

Packham, Chris. 2014. *The Wonder of Animals*. BBC TV.

Polanyi, Michael. 1974. *Personal Knowledge*. Routledge.
Sampson, Philip. 2018. *Animal Ethics and the Nonconformist Conscience*. Palgrave Macmillan.
Stone, Ken. 2018. *Reading the Hebrew Bible with Animal Studies*. Stanford: Stanford University Press.
Thomas, Keith. 1983. *Man and the Natural World*. London: Allen Lane.
Wordsworth, William. 1798. Lines Composed a Few Miles above Tintern Abbey. In *Lyrical Ballads*, eds. William Wordsworth and Samuel Taylor Coleridge. London: J. & A. Arch.
Wilkie, Rhoda M. 2010. Livestock/Deadstock. Philadelphia: Temple University Press.
Žižek, Slavoj. 2023. *Denazification Should Begin At Home, In Russia*. Interview with Vazha Tavberidze, 2.1.2023. https://www.rferl.org/a/zizek-interview-russia-denazification-ukraine-war/32204259.html. Accessed 5.9.2025.

CHAPTER 2

Speaking of Animals

Introduction

This chapter sets out a framework for exploring what can be said about animals. In particular, how different ways of speaking shape our perception of, and interaction with, them. It also introduces the nonconformist literature on which I extensively draw.

I introduce different ways of speaking, the way that they change over time, and their transformation in interaction with other discourses. The most influential contemporary discourses are, in one way or another, anthropocentric. I contrast this with the historical nonconformist discourse about animals which was distinctively theocentric. In that theocentric discourse the meaning of the lives of both animals and human beings is found in their ability to praise God, each in their own distinctively embodied way. Such a way of speaking gives animals their own voice; but that voice does not speak the language of the humans who have access to the resources animals need, and who have power over their bodies. This difference generates a duty upon humans to care for animals and to speak on their behalf.

P. J. Sampson, *Nonconformist Perspectives on Animals and Language*, The Palgrave Macmillan Animal Ethics Series,
https://doi.org/10.1007/978-3-032-12158-5_2

What Is It to Speak for?

To speak on behalf of other animals is to speak *for* them. Of course, animals speak for themselves, among themselves. Moreover, according to the nonconformists, they also speak beyond themselves; like all creation, they declare the glory of God. We need to speak for them only because we fail to hear their voice. Yet they do not speak into the void. The puritan John Trapp warns that we should show mercy, not tyranny, to "these dumb creatures"; dumb, that is, to us. But, cautions Trapp, beware, for animals are not dumb before God: if "we show not mercy", "he that hears the young ravens, may hear them", for "he is gracious" (Trapp 1660, vol. 3, 48).

But what is it to *speak for*? Perhaps, for their benefit. But then *we* judge the benefit, and we may too readily confuse it with our own. *Speaking for* must not make the *spoken for* voiceless in her own world (Deleuze and Guattari 1994, 109). As Trapp intimated, we should rather speak from their place, whence they address God as a speaking subject. But then we must first attend to their teaching. As the protestant reformer, John Calvin, observed, if we wish "to be considered a reasonable creature" we "would have to be sent to school to the beasts", so that they "should teach us what we must do" (Bouwsma, 1988, 140, 201). "[E]very creature" says the puritan Stephen Charnock (1864:1681, 27) "is our teacher, every part of the creature a lively instruction". Modern readers find nothing familiar in this way of speaking.

Ways of Speaking

The philosopher and Talmud specialist, Emmanuel Levinas (2003, 51), observes that the modern Western tradition of thought assumes that the world has an "autonomy and self-sufficiency of existence" which is usually taken to be anthropocentric. Most of the nonconformists discussed in this book would find this assumption deeply problematic. For them, existence is dependent, not self-sufficient or autonomous; far from being anthropocentric, the world speaks of something beyond itself. Indeed, they maintained that the world holds together only as praise and manifestation of the divine. Existence is, as it were, given as meaning; language is its common coin. Indeed, many adopted the long established tradition of speaking of the world as the *book* of creation, to be read (Tanzella-Nitti 2005).

This was not a matter of individual belief, any more than modern Western traditions of thought are individual beliefs. Rather, it was the ground of a communal discourse, and was articulated in confessions of faith, notably in the Westminster Confession of 1647.

Such communal discourses set out the way we talk about things. By articulating potentialities in the world, they profile our intentions and horizons of interest (Žižek 2008, xi). All thinking, in so far as it uses language, is built on them; they facilitate or inhibit what can be said, setting boundaries for the entities we talk about. Thus *animal* is a category we populate with entities generated from these discourses, depending upon whether we wish to admire them, pet them, experiment upon them, hunt them, or eat them. Communal discourses are produced and reproduced as collective responses to the challenges of everyday life, and are often so tacitly taken for granted as to be invisible.

Different Discourses, Different Meanings

Sometimes the boundaries between discourses are clear, as when we go to the butchers and buy meat. In England, if we intend to eat an animal, we speak of it in a culinary discourse and call it by a name of Norman French derivation: *mutton, pork, beef, capon*. Animal bodies so called are fungible, interchangeable commodities. Joan Dunayer (2001, 139, 137) considers language such as cutlet, pastrami, or sirloin to be "morally irresponsible because they obliterate the victims". She points to the animal industry's avoidance of the word *slaughter* in favour of *process, harvest*, or *go to market*.

As I have already noted, the Anglo-Saxon farmers, who cared for what the Norman conquerors ate, used a different discourse of husbandry care; they knew their animals individually, and called them sheep, pigs, cows, and chickens. The animals so named are non-fungible, each has its own needs; but neither are they unique; they are members of a species class.

Companion animals are not called by such a class concept, but by a proper name which is, like the companion which it names, unique. The Genesis creation account narrates that Adam named the animals brought to him (Gen. 2.18–20). Nowadays, this is often read through modern spectacles as Adam asserting the classifying power of anthropocentric modernity over other animals. But nonconformist commentators were not, in this sense, modern. They suggested that Adam gave each a name, that "harmonises with its nature"; an "appropriate", distinguishing name "agreeing with the nature of each" (Luther 1960:1538, vol. 1, 119. Trapp 1868:1657, Gen. 2.20. Calvin 1965:1554, 132). This suggests an intimacy of relationship based upon daily contact, not a species concept (Adam et al. 2019, 4.). Adam was not charged with naming plants, perhaps because plants are not individuals as animals are, but (almost) always species.

Sometimes boundaries blur. We rarely speak of animals with the consistent lack of feeling which we use of things. "Spasms of regard, tenderness, companionship" disrupt "unthinking callousness", as Mary Midgley (1996, 112) puts it. To prevent this while shopping, we use commodifying language. Farmers avoid giving personal names to "food animals" to avoid their companionship. One farmer described deliberately not thinking about what happens at slaughter in a "discourse of ignorance", made possible by the division of labour in the industry (Wilkie 2010, 153). The less we live in a mixed community with animals, the easier it becomes to maintain rigid boundaries; a failure to do so results in cognitive dissonance. Farmers and animal experimenters who work in mixed communities with animals employ a range of dissociative strategies to deal with dissonance (Joy and Tuider 2016, viii-xii).

Times Change

What can easily be said may change with time and place. For example, animal friendly language is more common now than in the past, as reflected in the vocabularies of welfare, legislation, and advocacy organisations.

Such changes can lead to misunderstandings. Thus, nonconformist discourse of the seventeenth century frequently uses *beast* rather than *animal*. The AV bible of 1611 reads *beast* (or its cognates) some 296 times; and *animal* not at all. *Beast* is currently a pejorative term, but seventeenth-century nonconformist usage should not be seen in similar terms. In that discourse, *beast* had much the same semantic range as the modern *animal*. Indeed, *beast* might actually have been *more* animal friendly, as it was commonplace to bring *beasts* into the home for protection, to share food, and so on. Nowadays such *animals* would mostly be housed outside the home in an industrial facility. By reading the nonconformist *beast* within a modern discourse, we change and even reverse its meaning. Similar considerations apply to bruite, inferior creatures, or worser creatures, now pejorative terms, but then encompassing simply the possession of less extensive attributes.

Sometimes, change can be radical, making one way of speaking incomprehensible to another. Examples of this will recur throughout this book. Indeed, it is a common experience among animal advocates when trying to explain their concern for, say, chickens, to those for whom "chicken" is merely an adjunct to "nugget", "wing", or "breast". It is to such discursive incomprehensibility that we turn next, mainly drawing examples from seventeenth-century puritans and dissenters.

Literal or Figurative

In seventeenth-century England, nonconformists found it easy to speak as though the world exists as praise and the manifestation of the divine. Most people today would find this much more difficult, including many Christian people. Thus, when animals and humans alike are called to praise God in Psalm 148, this is nowadays regarded as, at best, figurative in the case of animals, though not for humans. But this is to bring modern literary categories to bear on times when relationships between word and thing were in transition (Foucault 1966). There is no textual indication that, in Psalm 148, the animals are figures of speech whereas the people are the real article.

When Isaiah prophesied that a time will come when lions will eat straw (Is. 11.7, 65.25), modern readers see only figurative language. However, the congregationalist, Jeremiah Burroughes (1643, vol. 1, 578), read the pericope within the biblical arc of Creation, Fall, and Redemption. He saw a promise which "shall be literally fulfilled, and those other promises in Esay [Isaiah] and other places, where God says he will make *the Lion to eate straw with the Oxe,* and the like,... the creatures shall be brought into such a kinde of excellency in a manner as they were with Adam in Paradise". This does not imply that nonconformists were ignorant of figurative language when they read pericopes like these. Indeed, they were familiar with the intense debate in Western Europe about the relationship between word and thing, focussed around, but not limited to, the eucharistic elements of bread and wine. This debate brought a flexibility to pericopes such as Ps. 148 and Isaiah's prophecy. Burroughes, himself, was a graduate of Emmanuel College, Cambridge, and a member of the Westminster Assembly; his writings show a keen analytical mind, not given to naive errors of interpretation.

When God is said to provide food for the cattle and ravens when they "cry unto God", the modern reader naturally takes this as figurative (Ps. 147.9, Job 38.41; 39.3.). Yet John Calvin (1846, vol. 2, 1.18f) writing in the mid-sixteenth century, has a more nuanced reading: the young ravens, he says, cry to God in the *substantial* sense that "they confess by raising up their bills, that there is no supply for their want except God supports them". He considers that this "crying to God" is a "personification", but "was it not a calling on God as their nature admitted?". Calvin goes beyond metaphor by stopping short of it. "The natural cries of the distressed creatures are in substance nature's prayer to its Maker for relief and help", wrote Church of Scotland minister, David Dickson (1959:1655, 231, my emphasis).

Puritan, Joseph Caryl (1670, vol. 4, 208–9), was cautious of personifying animals, analysing such pericopes as instances of prosopopoeia and metonymy. Yet when the apostle Paul speaks of animals as waiting in "earnest expectation", Caryl clarifies this by making Paul's phrase refer to something more *substantial* and concrete, echoing Calvin's ravens "raising up their bills": "the creature stretcheth out (as it were) its head to look for these things, and is weary till its receives them". Presbyterian, John Flavel (1788, A2 and iv), wrote that the creatures speak to us "in a Metaphorical sense"; they "bear the figures and similitudes of many sublime and heavenly mysteries". Yet, this is not metaphor in the modern literary sense, but a *substantial* and engaged trope which is "represented to the senses" in everyday life. The Calvinist archbishop, George Abbot (1613, 454–5), observed that animals cannot call upon God with understanding such as men have, "and yet the Lord…respecteth all their cryes, and taketh many of those cryes as a kind of calling on him. We need not feare to speake that which the Spirit of God hath spoken". He considers that to treat this pericope as synecdoche is to be "more scrupulous then neede is". In nonconformist discourse, the language of "calling upon" does not require words "such as men have". Indeed, the language of "groans" which "far exceed the comprehension of our understanding" was already familiar to nonconformists from human calling upon God in prayer (Rom. 8.22–26. Calvin 1834:1539, 337). The animal's "expression of desire is a kind of prayer" said Calvinistic baptist, C. H. Spurgeon; "a kind of natural prayer to God, for relief", according to the leader of evangelical Methodism, John Wesley. Nonconformist, Matthew Poole, makes the comparison explicit: Their "roaring is a kind of natural prayer to God for relief, as the cries of infants are a kind of prayers to their mothers for the breast". If God listens to the prayers of animals, says Presbyterian, Matthew Henry (1710, Ps 104.19–30), "shall he not much more interpret favourably the language of grace in his own people, though it be weak and broken, groanings which cannot be uttered?".

This way of reading metaphor as *substantial* greatly declined from the late nineteenth century, but did not disappear altogether. Martyn Lloyd-Jones was a leading clergyman in the mid-twentieth-century revival of English Calvinism. Originally trained as a physician, he was not given to flights of fancy, yet he read in the nonconformist tradition. Lloyd-Jones agreed with Burroughes about Isaiah's prophecy of lions eating straw. He thought it *substantial*, writing that "the lion who is carnivorous now, will not be so [on the new earth]". This is not simply an idiosyncratic clerical opinion. F. F. Bruce, a distinguished Middle Eastern scholar, shared Lloyd-Jones' scepticism of unsubstantial figurative readings. He observed of Isaiah's prophetic "poetry" that it "enshrines no pathetic fallacy but something much more biblical and *substantial*" (Sargent, 1996, 153–4. my emphasis). There is a reality here about animals which metaphor tries, but fails, to grasp; a reality reduced by modern readings. It seems that, for nonconformist discourse, metaphor is less within the Isaiah pericope than the pericope is within metaphor.

Such radically different ways of speaking may understand the same "word" in different ways. In particular, a term used in one discourse may be taken up and transformed in another. Such examples of *appropriation* are common in religious discussions of animals as, for example, with the word "dominion". I will identify aspects of appropriation which will be important for our discussion below.

Appropriation

Sometimes one way of speaking may be appropriated by another, to serve an alien purpose. For example, the use of a *welfare language* on meat packaging may reassure consumers that animals are protected, while discouraging discussion of the cruelty entailed in breeding and killing them (Johnson 2012, 49). Appropriation here creates something new out of welfare language; namely the promotion of carnism. The plundered discourse is transformed, and acquires a new life which may be entirely alien to the original. Sometimes the novel transformation endures at the expense of the original. An obvious example is the appropriation of pagan festivals by the early christian church to constitute the church year. Another is the appropriation of nonconformist discourse by English Romanticism. Northcott (2018, 21–3) has noted a significant "Protestant genealogy to the ideas and experiences which led to the Romantic movement in Northern Europe". Romantic authors appropriated language from the nonconformist tradition of previous centuries, but now "detached from [its] religious provenance", as David Perkins (2003, 40) puts it, and integrated into the Romantic gaze. They created a distinctively new sensibility, alien to the original. In particular, the self-revelation of the creator God of nonconformity was transformed into the human discovery of sacredness imminent in an idealised Nature.

As well as transforming it into something entirely alien to the original, appropriating a way of speaking also points us to those features of the original that the appropriators "most wanted and most feared" (Brunstrom 2006, 158). As we shall see, this was the case with the dominant discourse's appropriation of ***dominion***, which effectively reversed its meaning in nonconformist discourse. It also marked the most feared feature of nonconformity; namely, that animals are not ours and it's not all about us.

Appropriation can enable us to read authors in a new way, and to group formerly disparate texts, creating relationships between them which were previously not apparent. Thus, animal ethics literature commonly speaks of a more humanitarian view of animals which emerged since ethics broke with the Christian church (Singer 2009, 186). This has produced new relationships between disparate texts as they are appropriated to create predecessors for animal friendly language. Authors as varied as Ovid, Seneca, Porphyry, Plutarch, de Montaigne, and Jeremy Bentham now acquire a new affinity as "animal friendly" (Singer 2009, 192–3, 203–4). By the same token, such appropriation marginalises the radically different philosophical orientations of these authors which were formerly centre-stage. It also excludes those authors associated with the pre-humanitarian break with the Christian Church. This can result in nonconformist literature being classified as preceding and resisting the modern humanitarian understanding, not as the result of the content of specific texts, but as a consequence of the way they are appropriated into a way of reading (Sampson 2018).

Having set out a framework to discuss discourses and their interactions, we now turn to specific ways of talking about animals.

Animal Discourses

When we talk about animals, we commonly use binaries to do so. The principle binary is human/animal, but there are many others: living/ non-living, companion/food, subject/object, seeing/seen, or speaking/ silent. There are a multiplicity of such binaries, each classifying animals in different ways, or for different purposes. Typically, each binary will lead to different, sometimes mutually incompatible, ways of speaking. For example, a farmer might speak of a lamb as ready for a succulent meal. However, every native English speaker knows not to ask how a companion animal would taste. As Ken Stone (2018, 79–80) has pointed out, there are also many different ways to speak of humans. As a result the binary human/animal is complex and liable to change over time.

Lisa Johnson (2012, 65, 83, 92) argues that one binary dominates contemporary talk about animals: the bifurcation into a dominant discourse which speaks of them as *property;* and a subordinate discourse of *living beings.* The dominant way of talking always trumps and often eclipses the subordinate one.

The Dominant Discourse

The dominant way of talking about animals as property tacitly relies upon a secondary binary division of the world into *property* and *owners*. This language of animals could as easily be speaking of tables or chairs. It is "being owned" which makes animals property, and in the modern world, the owner is almost always human. Property discourse may appear to be a way of talking about animals, but is actually not about animals at all, but about their owners; it is anthropocentric. Animals are consumer goods. The fact that owners are human does not imply that living property is always non-human. Human chattel enslavement has been common in many cultures, especially the early modern West. Women also have sometimes been regarded as the property of men.

Occasionally animals are regarded as owners but not within the dominant discourse. For example, in subordinate ecological discourses animals are said to *possess* their territories, not as individuals but collectively, and not by human statute but *naturally.* However, this subordinate discourse is trumped by private property relationships whenever humans wish to "develop" the land.

Johnson (2012, 69) notes that property discourse is closely linked to that of *instrumental use*, where humans are the users, and animals are the instruments used to achieve the desired human end; the human user is a subject, the animal used is firmly in the camp of object. Likewise in dominant discourse, animals are seen, not seeing (Derrida 2008, 4ff).

As a result of such links, the dominant discourse is best understood as a family of interconnected ways of speaking; animals as property certainly, but also for instrumental use, for consumption, as objects, as seen-not-seeing, homogenised not individual. All these ways of speaking tacitly posit a binary division of the world between humans and animals as "utterly different *kinds* of beings", with humans at the centre (Singer 2011,64). This anthropocentrism is culturally congenial, as it helps us evade the interests of animals in favour of our own. Despite criticisms, this dominant family of discourses retains a monopoly on symbolic capital when speaking of animals, creating socially legitimated meanings.

Subordinate Discourses

The Living Subject

Johnson identifies a more recent, subordinate way of talking about animals as "living beings". However, "living beings" could be either living *subjects,* or living things, *objects*, such as plants, or even rivers (Macfarlane 2025). Johnson, along with most animal ethicists, tacitly uses a discourse of living *subjects*, not objects; an animal's behaviour is a token that they have their own interests and are subjects of their own life. Like the dominant discourse, the subordinate one is a family of inter-related ways of speaking: animals are subjects, sentient, seeing, acting; they are ends-in-themselves, not for-the-use-of-others.

In the subordinate discourse of living *subjects*, humans and other animals occupy the same discursive space, and the binary human/animal becomes difficult to sustain as a difference of *kind*. It becomes one of degree: in complexity, language, tool making, and so on. Humans are an animal among animals, all members of the same Darwinian family. There is more, observed Singer (1999, 90) "to human existence than there is to bat existence", but they are both animals.

If animals are living subjects, they cannot easily be spoken of as objects or things; or as possessing properties such as being insensate, or being seen but not seeing; or as mechanisms obeying the laws of nature; or as being unworthy of ethical consideration. Thus, the view attributed to Descartes, that the cries of tortured animals were only the mechanical sound of a "little spring that had been activated", is not readily sayable within the subordinate discourse of living subjects (Rosenfield 1968, 54).[1]

By Comparison with Humans

Subordinate discourse has drawn upon a variety of comparisons with humans in order to argue that animals qualify for ethical consideration. Human capacities such as tool making, or the ability to reason or talk, once thought to distinguish humans from other animals, are seen as held in common. The discovery that virtually all animals possess the "neurological substrates that generate consciousness" in humans has bolstered the belief that animals also share the human experience of self-consciousness (Low 2012, 1–2). However, this sheds little light on whether human experience is a good guide to that of other animals. Mary Midgley (1996, 116) speaks of a sense that animals "mind what happens to them". But other animals might not "mind" the same things as humans.

The primary sense in humans is sight, and the iconic test of self-awareness is self-recognition in a mirror. But sight is an alien concept to bats, whose primary sense is echo-location. Roberto Cazzolla Gatti et al. (2015) suggest that a "sniff-test" is more appropriate than a mirror for dogs. The privileging of human capacities and experience by the subordinate discourse to argue that animals qualify for ethical consideration suggests a tacit, unrecognised anthropocentrism. This may be misleading in even so basic a matter as the expression of pain (Braithwaite and Droege, 2016). Prey animals commonly conceal pain to avoid being selectively predated; if the experience of pain is modelled on human expressivity, this can result in seriously underestimating animal suffering (Rollin 2011). It is anthropocentric to expect animals to show pain as humans do.

The reliance of subordinate discourse upon comparisons with human attributes to rank capacities, results in other animals struggling to perform, generates a hierarchy, and even excludes some altogether. If the

[1] Descartes may not himself have held this view (Cottingham 1993, 16–17).

standard were an ability to fly or to navigate in the dark, bats would rank far above humans, as most humans can't echo-locate at all.

The near universal practice of using human capacities as the standard of comparison renders subordinate discourse anthropocentric, albeit covertly so. But not just human capacities are privileged; the very construction of subordinate discourse entails the human ability to represent and symbolise. Only humans have the power to represent themselves, and their place among other creatures; to specify "what calls itself man and what he calls the animal" (Driscoll and Hoffmann 2018, 5). Thus, when Julian Huxley (1960, 253) spoke of man "linked by genetic continuity with all the other living inhabitants of his planet… they are all his cousins or remoter kin", he ostensibly presented a more humble, less anthropocentric view than the dominant discourse. This is what Žižek (1999, 132–33) calls "false humility", one which pretends to speak for all animals, and not just its own interests. Portrayed creatures are trapped in their representation as surely as in a snare; they await humans to pronounce their fate. Representation is the anthropocentric denial of voice, draining animals of agency. The current "threshold requirement for participation in political discourse" is this human voice; this is simply because humans have set the threshold, ensuring that other voices are silent. "Consequently animal interests are generally ignored or not understood" in policy development (Johnson 2012, 24). Nonconformist discourse's reference beyond itself cuts into and disturbs a "true arrogance" posing as a pretended humility (Žižek, 1999, 132–33). In nonconformist discourse, animals are not primarily represented as much as they are embodied (Falque 2016).

The tacit anthropocentrism of the subordinate discourse of living beings has practical consequences for non-human animals which are not always recognised.

Jeremy Bentham's Diet

In the animal ethics literature, the remark by Jeremy Bentham (1970, 282–83) in 1789 that: "the question is not, Can they *reason*? nor, Can they *talk*? but, Can they *suffer*?" is widely seen as a turning point by foregrounding sentience and prioritising a capacity to suffer over, say, reasoning skills. Indeed, Singer (2009, 207) regards it as inaugurating a new era. However, the quotation concerned comes from a footnote which also contains the rarely cited observation that "[i]f the being eaten were all, there is very good reason why we should be suffered to eat such of them [animals] as we like to eat: we are the better for it, and they are never the worse".

It would be difficult to think of a more anthropocentric logic than: if we "like to eat" animals, that is a "very good reason" to do so. Because human experience is its touchstone, Bentham's discourse of the living subject collapses into the dominant discourse as soon as an animal's interest in *living* meets the human interest in *eating*. Its covert anthropocentrism makes it impossible to conceive that animals might be "the worse" for being dead. Bentham recognised the importance of suffering. But his representation of animals traps them in his footnote with the intention of eating them because he had a mind to. Nothing says I have power over your body so much as "I will eat you". Bentham's utilitarian text has the final say over his footnote on suffering. The fact that the full footnote is so rarely cited obscures the consequences of representing animals by human comparisons (Boddice 2008, 126–27). Both dominant and subordinate discourses hang upon humans, their capacities and attributes, albeit the one explicitly and the other implicitly.

Living Objects

Although animal ethics tacitly identifies living *beings* with living subjects, they could also be living *things* or *objects*.

Discourses of living *objects* are common is botany, ecology, and related fields, where the emphasis falls upon plants or species. However, it has also been used of animals. The best known instance is the mechanical philosophy attributed to Descartes and mentioned above. Nowadays, the discourse of living objects is most often found in the meat industry. "To visit a modern CAFO (Confined Animal Feeding Operation)", wrote Michael Pollan (2002), "is to enter a world that, for all its technological sophistication, is still designed according to Cartesian principles: animals are machines incapable of feeling pain".

Violence and Cruelty

There are two fundamental ways to be violent towards animals: symbolic violence; and manifest violence. These are often interconnected, each facilitating the other. The symbolic violence of the dominant discourse makes it easier to slaughter, and the manifest violence of slaughter reinforces dominant discourse.

A basic requirement for the manifest violence of cruelty is that the animal is able to suffer. Since neither insensate objects or mechanical devices can suffer, both the dominant discourse of property and that of living objects make it hard to discuss cruelty (Midgley 1996, 112–114). Symbolic violence forecloses the human ability to speak of cruelty. It becomes congenial not to hear them; they are rendered voiceless. Subordinate discourse has its own subtler symbolic violence in this respect. It may speak of living subjects, but its covert anthropocentrism makes animals mime voices actually spoken by humans. In neither discourse, do animals have a voice with which to speak for themselves. Symbolic violence also blinds us to the animal's tormented gaze.

Claude Bernard was notorious for his brutal vivisections in nineteenth-century France. For him the ideal vivisector is a scientist who "no longer hears the animals' cries of pain, and is blind to the blood that flows, and who sees nothing but his idea and organisms that conceal from him the secret he is resolved to discover" (Preece 2002, 309).

For Bernard, the living object can neither cry in agony, nor be seen. It is fatal to such a way of speaking if the living *object* becomes a *subject.* One way that this could happen is to see a living subject looking back at us. As Derrida notes, in discourse of the living *object,* people have "seen, observed, analyzed, reflected on the animal, but ... have never seen themselves being seen by the animal" (Derrida 2008, 13). The animal is always already the object of observation; "the fact that they can observe us has lost all significance" (Berger 1991, 16). When the tormented animal looks at us, we look into her eye, and see only a reflection of ourselves. For Alfred Wallace (1889, 37), the "supposed "torments' and 'miseries' of animals... are the reflections of the imagined sensations of cultivated men and women in similar circumstances". Most people not in urgent need of therapy find this a troubling way of speaking.

Sometimes the symbolic violence which robs the animals of their gaze may become unstable under the impact of manifest violence. Rhoda Wilkie (2010, 163, 16) comments that the most difficult part of a slaughterer's job is shooting the animal face to face when he cannot avoid her "moon-eyes look..., a bad, longing, unsure, looking-for-comfort [look]". "You see that walking towards you and you've got to shoot it". As one ex-beef farmer remarked: "We did [our] best to look after them [the cattle], but you knew you were going to betray them. You really couldn't look them in the eye" (Sandhu 2018, 15).

Occasionally, a discourse of the living object can suffer catastrophic collapse. Charles Darwin (1871, vol. 1, 40) writes of a "dog suffering under vivisection, who licked the hand" of the vivisector; in the dog's act, the discourse of the experimental living object collapsed into one of a living subject. This man, wrote Darwin, "unless he had a heart of stone, must have felt remorse to the last hour of his life".

Language which robs animals of their voice and their gaze is entirely foreign to nonconformist discourse. For John Wesley (1958:1872, 343; also vol. 8, 496), animals are not "mere machines", they have pleasure or pain, see and hear, taste and smell, remember, and move. The nonconformist, John Ray (1743, p54-5, 176), argued against mechanism and for the reality of animal suffering, on the basis that the bible condemns cruelty which would make no sense if animals were machines. Similarly, they found in the biblical texts animals which return our gaze; this places animals in the position usually occupied exclusively by man (Driscoll and Hoffmann 2018, 2). Puritan member of the Westminster Assembly, George Walker (1641, 210), speaks of a mutual joy of the animals and humans at their creation, precisely because they encounter one-another in a mutual gaze: "to all living creatures they [Adam and Eve] appeared lovely, and full of beauty, and majestie. It was the creatures delight to see them, and to looke on them; and it was their [Adam and Eve's] joy to see the creatures admiring them, and rejoicing in their sight and presence". This is a language more commonly used of lovers than of machines.

I have discussed both the structure and interactions of discourses, and have given examples of actual ways of talking about animals. But discourses do not simply arise, they are constructed from what preceded them. Key instances of this will be important for our discussion.

Constructing Discourses

In the contemporary West, secular humanist ways of speaking hold a monopoly on symbolic capital, and create socially legitimated meanings (Bourdieu 2005). Both dominant and subordinate discourses are usually constructed within this frame.

Many texts on animal and environmental ethics, present two related narratives. The first finds the roots of the dominant discourse in Christianity, understood as anthropocentric. The second sees the Enlightenment, with its factotum Darwinism, as the origin of the more recent subordinate discourse. Both narratives appropriate *dominion* and *imago dei*, transforming them in distinctive ways.

The Appropriation of Dominion

The source of the dominant discourse is commonly located in Genesis (1.26–8) where God gives "man" "dominion" over the animals. Ethicist, Peter Singer (1975, 226), is blunt but not atypical: 'It can no longer be maintained by anyone but a religious fanatic that man is the special darling of the whole universe, or that other animals were created to provide us with food…". In a much cited article about environmental degradation, the historian, Lynn White (1967, 1205), writes that "Christianity is the most anthropocentric religion the world has seen" as "no item in the physical creation had any purpose save to serve man's purposes". For George Hendry (1980, 17), Luther's doctrine of creation "reduced the whole world of nature to a repository of goods for the service of man". As Keith Thomas (1983, 17, 151) has documented, there are many examples of Tudor and Stuart theologians doing precisely this. He suggests that they tailored their interpretation of the bible to the "long established view that the world had been created for man's sake" which "underlay the actions of that vast majority of men who never paused to reflect upon the matter". This appropriation of the biblical text transformed it into something the nonconformists would not have recognised. It also points to that feature of nonconformist discourse which the dominant discourse most fears; that animals are not ours, and it's not all about us. This appropriation of dominion contrasts sharply with nonconformist discourse in three principal ways.

Firstly, for nonconformists, animals are not human property. S G De Graaf (1977, 34) summarises their view: "There was nothing man could call his private property; everything belonged to God, who had made it. Man had no natural rights given with his nature…". Secondly, they overwhelmingly rejected the view that "other animals were created to provide us with food"; dominion is not carnism. Indeed, the verses following the often cited grant of dominion (Gen. 1.29–30) allocate a vegetable diet to humans, as well as to every bird and land animal. Finally, and linked to the denial that creation is human property, animals are not solely "to serve man's purposes" (White op cit.). Rather, their purpose, enshrined in the Westminster Confession (1647, IV.i and V.i), is to praise and manifest God.

Santmire (1985, 124–133) illustrates this appropriation of nonconformist features in the case of the Magisterial Reformers, Martin Luther and John Calvin. He notes Calvin's "exuberant" praise of "nature's beauty in itself" and remarks Luther's "almost Franciscan" approach; neither, he says, considered "that the only divine raison d'etre for the whole creation is the redemption of the human creature". Despite this, they have been read within a tradition which transformed them into precursors of the dominant discourse. "[T]rapped in a soteriological tradition inevitably putting man at the centre", their theology "can appear to be radically anthropocentric". In this book, I will endeavour to free their theology from this trap, and illustrate the enduring value of the nonconformist discourse which drew so heavily on the original insights of Luther and Calvin.

The Appropriation of the Imago Dei

By contrast with the origin of the dominant discourse, contemporary narratives of the origin of subordinate discourse focus on the break with the Christian church epitomised by the Enlightenment and the Darwinian tradition. Singer (2009, 186, 203) nicely summarises this narrative. A "more enlightened view" of animals emerged, he says, as thinkers broke free of the church, culminating in the Enlightenment which produced "a newer, more benevolent attitude". In particular, it brought recognition that animals are living subjects who can suffer, finding its clearest expression in Bentham's famous footnote. But, says Singer (20,011, 64; 2009, 206), the real turning point was "Darwin's discovery of our origins and the associated decline in the credibility of the story of our Divine Creation in the image of God. ... the difference between us and animals is one of degree, rather than of kind....". With the acceptance of Darwin's theory, we reach a "modern understanding of nature".

Dominion and *Image of God* have here been appropriated as a foil for "a newer, more benevolent attitude". As Midgley (2013, 50; 1996, 103) observes, the Enlightenment break with religion transformed the Image of God into an "over-confident, unselfcritical pretension to 'objectivity'", reinforced by a vision of human rationality dominating the natural world. Once free from religious constraint, she claims, human "grandiosity became quite uncontrolled". The anthropocentric dominant discourse was born, and replaced God with man in what she terms *anthropolatry*, self-centring man's self-idolisation. Subordinate discourse is constituted by dis-identifying with this transformed imago dei to constitute a more modest and benevolent estimation of man's relationship with animals.

What of the claim that Darwinism played a pivotal role in the origin of the subordinate discourse by establishing that, contrary to the bible, "the difference between us and animals is one of degree, rather than of kind"? Richard Ryder (2016) puts it concisely: "Charles Darwin told us that we are all animals. These are our relatives. We are related through kinship" and are therefore "in the same moral community as ourselves". We are all, as it were, members of the same family.

This account of the "family" metaphor relies upon the discursive norms of human kinship prevalent in Victorian patriarchy: a "responsibility" to protect dependents. But the norms of *Darwinian* kinship are more like those of Victorian capitalism: competition and the survival of the fittest. For the anti-vivisection campaigner, Frances Power Cobbe, this fatally undermined the *purported logic* of Darwinian influence. Puzzled by Darwin's own support for vivisection, she lamented that his disciples were leading vivisectionists, advocating a "moral theory... that the weak have absolutely no claims at all against the strong, but may be tortured ad infinitum even on the chance of discovering something interesting to the lordlier race" (Hamilton 2004, vol. 1, 103–4). This is not the "more benevolent attitude" hoped for. One young vegetarian abandoned her diet on learning about Darwinism, remarking: "Well, in the natural world there's this thing called survival of the fittest and it's OK to eat animals" (Singer and Mason 2006, 186).

Transcendence

A striking feature of these narratives of the origin of dominant and subordinate discourses is the role of religion, specifically Christianity. In this foregrounding of the break with Christianity, the transcendent is cancelled, crossed out, repressed. If used at all, "religious" language provides a historical contrast to modern, humanist understandings which are more animal friendly. Emmanuel Levinas (2003, 51) argues that Western thought's abandonment of its "historical concern with transcendence" for an anthropocentric "autonomy and self-sufficiency of existence" has not gone well. Historically, the repressed transcendent did not long remained barred, but repeatedly returned, invading secular debate in the form of *Reason* and *Humanity*.

This was most obvious in the appropriation of religious awe by the Romantic movement, sublated, preserved, and transformed into the sublime, the ineffably boundless, the "absolutely great", to be found especially in nature (Kant 1951, para 23, 25, 27). But the return of transcendence was also apparent in politics and the human sciences. It returned in the cult of *Reason* and of the *Supreme Being* associated with the French Revolution. Auguste Comte, pioneer of positivism and scourge of religion as pre-scientific, created a Religion of Humanity, "a complete system of belief and ritual, with liturgy and sacraments, priesthood and pontiff, all organised around the public veneration of Humanity" (Davies 2008, 28–9). The elevation of humanity to transcend nature inevitably demoted other animals to a radically inferior status.

Bruno Latour (1993, 33) argues that transcendence has always resisted banishment, remaining concealed in the heart of modernity. Even as we took ourselves to be on the other side of a great rupture with religion, we behaved otherwise. A "crossed out God" acts as guarantor for the twin pillars of science and humanity which have never seen themselves as hanging in the void. Both reserve the possibility of appealing to transcendence if necessary. We continue to have confidence that we have not constructed reality, that it transcends us; "the truth is out there", to be discovered. Even as the soul was banished from psychology, we continued to have faith in-humanity, to live as though the human spirit could inspire a better world (Condorcet 1795). Again, the human pretension to transcendence is bad news for other animals.

By contrast with discourses which bar God only to have transcendence return, the nonconformist way of speaking begins with transcendence, with creation pointing beyond its material conditions as meaning. God, the radical denial of human "autonomy and self-sufficiency of existence", is at the centre, eluding integration into knowledge other than by revelation (Levinas 2003, 51; 1969, 40). God addresses us before we can speak for ourselves. Moreover, pace White's (1967, 1205–6) claim that Christianity has desacralised the world, "If the fulness of the earth is the Lord's, there is nothing in the world that is not sacred and pure" (Calvin 1848:1546, vol. 1, 344). This theocentric discourse puts all animals, human or not, in the same boat as creatures, manifesting and praising God, but not themselves transcendent. Ironically, the explicit recognition of God's transcendence results in a foreclosure of the radical superiority of humans over other animals.

Having discussed the appropriation of nonconformist discourse, and given examples of its transformation into something alien to the original, the time has come to introduce those aspects of nonconformist discourse which will be developed in succeeding chapters.

In addition to dominant and subordinate discourses, Johnson also identifies *subjugated* discourses, those once authorised but now buried or disqualified. I have argued elsewhere that nonconformist discourse was once authorised by a significant religious community to disrupt the then dominant ways of talking about animals, both religious and secular. It generated new forms of agency and institutions, warranting and obtaining an authorised place in the public domain. However, its incisions into dominant ways of talking about animals were progressively sutured in modernity. By the early twentieth century, it had faded as a cultural force, and is now subjugated in contemporary debate, despite vestiges that remain in our way of talking about animals (Sampson 2018, ch. 8). Here, I will highlight only those features which disrupt the dominant and subordinate discourses of today, and which will be developed further in following chapters.

Nonconformist Discourse

Until the end of the nineteenth century, a now subjugated nonconformist discourse permitted animals to speak for themselves. It included the everyday experience of animals, often clothed with biblical allusions which found multiple points of resistance to disrupt the then dominant ways of talking about them.

Recovered Readings

As I have illustrated above, when read within the contemporary secular discourse of human self-sufficiency, nonconformist texts are appropriated and transformed. Recovering them also recovers the points of resistance they offered to the dominant discourse of their day. Seven recovered readings will be important for the chapters which follow. I will briefly outline each; a full discussion will be found in subsequent chapters as need arises.

Theocentrism, Not Anthropocentrism

A fundamental distinction between nonconformist discourse and that of the modern secular world is the pivot on which each revolves. Both the dominant and subordinate discourses about animals are anthropocentric, albeit overtly and covertly so, respectively; humans acquire a tacitly transcendent status over other animals. Nonconformist discourse is theocentric, demoting humans from the centre of things; humans and other animals alike are creatures who work in mutuality and difference to the praise of God and the good of all. The twentieth-century Calvinist, Benjamin Warfield (1927, vol. 5, 358), emphasised that theocentric discourse "withdraws the eye from the soul and its destiny, and fixes it on God and His glory. ... its highest zeal is for the honour of God, and it is this that quickens its emotions and vitalizes its efforts. ...the end of human as of all other existence, ... is the glory of the Lord of all... It begins, it centres and it ends with the vision of God in His glory".

Meaning, Not Essence

Western thought has consistently privileged autonomous being, the presence of that which *is*, over that which allows presence itself to be possible. Nonconformist theocentrism rejects the autonomy of being, making a radical distinction between Creator and creation. The implied ontology came to maturity in the work of the twentieth-century philosopher, Herman Dooyeweerd (1969, vol. 1, 4), for whom: "Meaning is the being of all that is created and the nature even of our selfhood". Meaning here is not simply linguistic reference, but is given as creation; in the case of animals, as embodied. Moreover, meaning is not free-floating but directed; it manifests and glorifies the Creator, finding its pleasure and fulfilment in praise.

Insofar as language has to do with how meaning is made, it represents the common coin of creation. "In the beginning was the word"; "God said, Let there be... and it was so" (John 1.1; Gen.1.6–7). Sharing something of the theologian Catherine Pickstock's (1998, xiii) doxological view, for nonconformists, "language exists primarily, and in the end only has meaning as, the praise of the divine". As embodied meaning, animals are portrayed realistically and sympathetically, not idealised or enfabled; yet as *embodied* meaning, they can praise without being reduced to metaphor (Perkins 2003, 47–8).

Dependence, Not Self-sufficiency

Dominant and subordinate ways of talking about animals are either explicitly or tacitly anthropocentric. Nonconformist discourse radically rejects this, maintaining that created existence is dependent, and holds together only as praise and manifestation of the divine. Humans and animals share agency and vulnerability in their common task (Stone 2018, 146).

Care, Not Instrumentalism

The dominant discourse of animals is explicitly instrumental. Subordinate discourse is indirectly instrumental in that it tends to favour disengagement between humans and other animals in order to avoid instrumentality (Seidenberg. 2019, 161–62). Neither provides resources for reading "use", otherwise than instrumentally. Thus, when Calvin (1845:1559, I.14.22) says that God "created all things for man's sake", or George Abbot (1613, 450) that animals are "subjected to [man's] use", their language is now read as modern instrumentalism, commodifying animals as a means to man's end. This enables the confident assertion that, according to the doctrine of dominion in the Bible, "God has created the World... [and chose] to license Adam and Eve to do what they liked with it..." (Toynbee 1974, 141). I will discuss this further in Chapter 4, but we might get closer to the nonconformist sense of dominion by thinking more of a sense of gratitude, identity, and collective solidarity with creation, than of instrumental commodification.

Common Use, Not Property

As we have seen, the dominant discourse of property ownership claims to trace its origin to Christian theology, but its language is more recent, and was constructed from that of maturing capitalism. By contrast, early modern property relationships allowed for a variety of economic conventions, including that of the steward, villicus, bayliff, farmer with usufruct, and various common rights such as turbary, pannage, or estover. The teaching of the medieval church included a strong theme that property claims are vacuous in themselves, subject to proper use for the benefit of others, and are invalidated when used for one's own desires. The nonconformists wrote against this background, and could scarcely miss the biblical assertion that "the earth is the Lord's", repeated more than ten times. God may have "lent [animals] to us for our help and service... besides the delight that many of them ministere to us...", as the puritans, John Dod and Robert Cleaver (1612, 141) say, but we have no natural ownership rights over them given with our nature (De Graaf 1977, 34).

If we abuse them, we show our ingratitude, and invite God's protective displeasure.

Goodness, Not Tyranny

When appropriated by the dominant discourse, *dominion* is transformed it into something distinctively modern. Singer (2009, 187, 203–4) is not unrepresentative in describing the *dominion* of vegan Eden as a "benevolent despotism"; he considers that "Bentham was perhaps the first to denounce 'man's dominion' as tyranny rather than legitimate government". Yet nonconformist discourse denounced tyranny over animals long before Bentham. Matthew Henry (2008:1710, Num. 22.30—my emphasis) argued that the "dominion God has given us over the creatures is a good reason why we should *not* abuse them. We are their lords, and *therefore* must not be tyrants". Trapp (1660, vol. 3, 48) warned that God will "not have tyranny exercised" on animals, and puritan, Nathaniel Ward (1641, para. 92), made it a "liberty" of the Massachusetts Bay Colony that "No man shall exercise any Tirranny or Crueltie towards any bruite Creature which are usuallie kept for man's use". It is worth noting that "tyranny" is a very strong term to use; even more so for nonconformists, many of whom had experienced oppression from Church or State, and some had lived through a civil war fought to oppose tyranny. To apply this word to animals is to draw an equivalence with human experience. The appropriation of dominion as tyranny has effectively reversed its nonconformist emancipatory significance.

Rights and Justice, Not Exploitation

Within the Enlightenment tradition, justice speaks the language of rights. The Enlightenment philosopher, Immanuel Kant (2001, 212), had tethered rights to humans as ends-in-themselves, and concluded that we owe animals "no direct duties". This remains the case for the dominant discourse, but subordinate discourse has recently extended rights to animals.

Although now little known, a language of rightful dealing and duties preceded the Enlightenment in nonconformist discourse (Sampson 2018, 94f). This was based, not on attributes derived from human *being* and extended to other animals, but on shared creatureliness. This anticipation of "theos rights" was tethered to keeping God's covenant, not to human being (Linzey 1989, 95, 97). Thus Calvin (1847:1563, vol. 3, Deut. 25.4) wrote that, in the bible, "men are so instructed in equity, that they are bound to exercise it even towards the brute animals".

Secular appropriations of this language simultaneously detached it from both its nonconformist provenance of creatureliness, and from non-human animals, to collapse into modern rights language reserved for political debate about enfranchised men (Northcott 1996, 97). For animals, the language of rightful dealing and justice became largely silent until the late twentieth century. Its recovery has been one-dimensional, lacking as Linzey (1989 p95) points out, a language of mercy, compassion, respect, care, self-sacrifice, even of love, as well as justice. The nonconformist norm for human interactions with animals is love, not despotism. We should care, say Dod and Cleaver (1612, 141), not out of self interest, but for "love to the creature, or him that made it". In nonconformist discourse, belonging is not necessarily a property relation, a gift is not entitlement, and dominion is not tyranny.

This chapter concludes by returning to the questions with which we started: how do animals speak; and how might we speak *for* them?

Not Speechless, Not Silent

The dominant discourse of modernity renders animals silent; to speak is to become either a "not-animal", or a fabulous, mythical talking creature. In the subordinate discourse, animals struggle to speak, to break free of miming to a human script. By contrast, the nonconformist emphasis upon the *meaning* rather than the *being* of creation makes animals inherently expressive. Animals praise and manifest God; they cry to God, and He hears them. They have a voice because they have standing; as a creature of God, they have an office of authority to speak. For Calvin (1583, 877), if a cruel man says, "tush, I care not, for it is but a brute beast: [I answer again]: Yea, but it is a creature of God". So beware, cautions Trapp (1660, 48), if "we show not mercy" for they are not dumb before God.

According to Their Kind

Of course, nonconformist discourse recognised that animals do not speak as we do, but according to their kind. Their voice is one of immediacy and connection which rests in the fissures, subjectivity, and complex prosody of language: the rhythm and musicality of bird song; the insistent tone of a plaint; the sharp intonation of an agonised cry. It may also be a symbolic field of meaning which we cannot yet understand (Linzey 2009, 41). The animal can look at us and, "in a word, without a word, address" us (Driscoll and Hoffmann 2018, 2). In nonconformist discourse, animals are neither symbol, nor allegory, nor foil for what it is to be human; but flesh and blood, embodied meaning, declaring their creator. The cries of such creatures change us. As philosophers, Deleuze and Guattari (1994, 109), observe: the "agony of a rat or the slaughter of a calf remains present in thought not through pity but as the zone of exchange between man and animal in which something of one passes into the other". This may not be congenial to humans.

The creatures, says Calvin (1845:1559, 1.5.15) bear a"magnificent testimony… to the glory of their Creator", yet we decline to hear. "It were, indeed, a strange defence for man to pretend that he has no ears to hear the truth, while dumb creatures have voices loud enough to declare it; … to excuse himself on the ground of weakness of mind, while all creatures without reason are able to teach". Matthew Henry complains, "We hear it not, but thus the whole creation groans, being burdened".

As an animal's voice is not modelled on human capacities or attributes, it is not simply an echo. They are, as Calvin noted, "able to teach". But we may not wish to hear. Matthew Henry (2008:1710, Num. 22.23–33) noted that Balaam's donkey several times turned aside to warn Balaam of a danger on the path; she spoke but he did not hear. The vivisected dog who licked the hand of her tormentor was surely eloquent, but was rebuffed.

Speaking for Animals

Matthew Henry (ibid), speaking of a beaten donkey, argues that God requires advocacy: "The righteous God will not see the meanest and weakest abused; but either they shall be enabled to speak in their own defence or he will some way or other speak for them". "We cannot open the mouth of the dumb, as God did" for Balaam's donkey, "but we may and must open our mouth for the dumb". Indeed, the very lack of *human* speech itself spoke of a duty to enter the non-human world, and to listen. "A beast" advises Calvin (1583, 770), "cannot speak to move us to pitie and compassion; and therefore we must go to him of our own good will, though we be not moved or requested thereunto". Anglican clergyman, James Granger (1774, 10), advocated for animals from the pulpit, since they "cannot speak for themselves". The lack of a language to give consent, observes Linzey (2009, 22), implies that our responsibility to consult is the greater. For nonconformists, this duty refers, not to attributes derived from human *being*, but to a shared creatureliness, each serving God differently, according to its kind. Evangelical calvinist, Richard Hill, opposing bull baiting in 1802 said he spoke on "behalf of a race of poor friendless beings who certainly cannot speak for themselves". In this, the voice of those nonconformists echoes that of recent authors for whom animals have indeed spoken, albeit differently from humans; and who advocate a practice of attentive listening, of "listening otherwise" (Driscoll and Hoffmann 2018, 3).

From Speech to Song

According to the Westminster Confession (1647, IV.i and V.i), animals were created to praise and manifest God. This is the form and content of their voice: they give, as Calvin put it, "magnificent testimony" to the glory of God. The next chapter will explore the manner of this "testimony", their longing to manifest wonder and joy. I shall develop the nonconformist understanding of their voice, not primarily as "speech", but as "song". Subsequent chapters consider the interactions between animals when "singing together", the distinctive human calling to facilitate their praise, and the consequences of animals being "unwillingly" subjected to "groaning in pain", yet hoping for restoration (Rom. 8.18–25).

References

Abbot, George. 1613. *An Exposition upon the Prophet Jonah*. London: Richard Field.

Adam, Margaret B., David L. Clough, and David Grumett. 2019. A Christian Case for Farmed Animal Welfare. *Animals* 9: 1–13.

Bentham, Jeremy. 1970. *An Introduction to the Principles of Morals and Legislation*. London: Athlone.

Boddice, Rob. 2008. *A History of Attitudes and Behaviours toward Animals in Eighteenth- and Nineteenth-Century Britain*. Lampeter: Edwin Mellen Press.

Bourdieu, Pierre. 2005. *Language and Symbolic Power*. Cambridge: Polity.

Bouwsma, William J. 1988. *John Calvin*. Oxford: OUP.

Braithwaite, Victoria A. and Paula Droege. 2016. Why Human Pain Can't Tell us Whether Fish Feel Pain. *Animal Sentience* 3 (3).

Brunstrom, Conrad. 2006. Leaving the Herd': How Queer Was Cowper? *British Journal for Eighteenth-Century Studies* 29 (2): 157–167.

Burroughes, Jeremiah. 1643. *An Exposition of the Prophesie of Hosea,* vol. 1, 578. London: W. E. and I. G. for R. Dawlman.

Calvin, John. 1583. *Sermons on Deuteronomy* Trans. Arthur Golding. London: John Harrison 1583.

Calvin, John. 1834:1539. *Commentary upon Romans.* London: I. B. Seeley.

Calvin, John. 1845:1559 *The Institutes of the Christian Religion.* Trans. Henry Beveridge. Edinburgh: Calvin Translation Society.

Calvin, John. 1846. Commentary on Joel. In *Commentaries on the Twelve Minor Prophets.* Edinburgh: CTS.

Calvin, John. 1847:1563. *Commentaries on the Four Last Books of Moses.* Edinburgh: CTS.

Calvin, John. 1848:1546. *Commentary on Corinthians* Edinburgh: CTS.

Calvin, John. 1965:1554. *A Commentary on Genesis.* London: Banner of Truth.

Caryl, Joseph. 1670. *An Exposition with Practical Observations upon the Book of Job.* London: Thomas Sawbridge.

Gatti, Cazzolla, Alena Velichevskaya Roberto, Benjamin Gottesman, and Karen Davis. 2020. Grey Wolf May Show Signs of Self-Awareness with the Sniff Test of Self-Recognition. *Ethology Ecology & Evolution* 33 (4): 444–467.

Charnock, Stephen. 1864:1681. Discourse on God's Wisdom. In *Works,* vol. 2. Edinburgh: James Nichol.

Condorcet, J. de Caritat. 1795. *Outlines of an Historical View of the Progress of the Human Mind.* London: J. Johnson.

Cottingham, John. 1993. *A Descartes Dictionary.* Oxford: Blackwell.

Darwin, Charles. 1871. *The Descent of Man.* London: John Murray.

Davies, Tony. 2008. *Humanism.* London: Routledge.

De Graf, S. G. 1977. *Promise and Deliverance* New Jersey: Presbyterian and Reformed.

Deleuze, Gilles, and Félix. Guattari. 1994. *What Is Philosophy?* New York: Columbia University Press.

Derrida, Jacques. 2008. *The Animal that Therefore I Am* Ed. Marie-Louis Mallet. Trans. David Wills. New York: Fordham University Press.

Descartes, Renee Discours vol1 pp 115–118 in Andrew Linzey & Tom Reagan (eds) (1989) Animals and Christianity: A Book of Readings SPCK.

Dickson, David. 1959:1655. A Commentary on the Psalms London: Banner of Truth.

Dod, John, and Robert Cleaver. 1612. *A Plaine and Familiar exposition of the eleventh and twelfth chapter of the proverbs.* London: Thomas Man.

Dooyeweerd, Herman. 1969. A New Critique of Theoretical Thought. Presbyterian and Reformed.

Driscoll, K., and E. Hoffmann. 2018. Introduction. In *What Is Zoopoetics?,* ed. K. Driscoll and E. Hoffmann, 1–13. Palgrave Macmillan.

Dunayer, Joan. 2001. *Animal Equality: Language and Liberation.* Derwood, Maryland: Ryce Pub.
Falque, Emmanuel. 2016. *Crossing the Rubicon.* New York: Fordham University Press.
Flavel, John. 1788. *Husbandry Spiritualised.* London: John Binns.
Foucault, Michel. 1966. *Les mots et les choses.* Paris: Gallimard.
Granger, James. 1774. *An Apology for the Brute Creation.* London: T. Davies.
Hamilton, Susan, ed. 2004. *Animal Welfare & Anti-Vivisection 1870–1910.* London: Routledge.
Hendry, George. 1980. *Theology of Nature.* London: Westminster Press.
Henry, Matthew. 2008:1710. *Commentary on the Whole Bible.* Peabody, MA: Hendickson.
Huxley, Julian. 1960. The Evolutionary Vision. In *Evolution after Darwin,* ed. Sol Tax and Charles Callender, vol. 2, 249–62. Chicago University Press.
Johnson, Lisa. 2012. *Power, Knowledge.* Animals: Palgrave Macmillan.
Joy, Melanie and Jens Tuider. 2016. The Rise of Veganism. In *Critical Perspectives on Veganism,* ed. Jodey Castricano and Rasmus R. Simonsen, v-xv. Palgrave Macmillan.
Kant, Immanuel. 1951. *Critique of Judgment.* Trans. J.H. Bernard. London: Macmillan.
Kant, Immanuel. 2001. *Lectures on Ethics.* ed. P. Heath and J. B. Schneewind. Trans. P. Heath. Cambridge: CUP.
Latour, Bruno. 1993. *We Have Never Been Modern.* Cambridge Massachusetts: Harvard University Press.
Levinas, Emmanuel. 1969. *Totality and Infinity.* Pittsburgh: Duquesne University Press.
Levinas, Emmanuel. 2003. *On Escape.* Stanford University Press.
Lewis, C. S. 2009. *The Magician's Nephew.* London: Harper Collins.
Linzey, Andrew. 1989. *Christianity and the Rights of Animals.* New York: Crossroad.
Linzey, Andrew. 2009. *Why Animal Suffering Matters.* Oxford: OUP.
Low, P. 2012. *The Cambridge Declaration on Consciousness.* Proceedings of the Francis Crick Memorial Conference, Cambridge University, July 7 2012.
Luther, Martin. 1960:1538. *Works.* Ed. Jaroslav Pelikan. Saint Louis: Concordia.
Macfarlane, Robert. 2025. *Is a River Alive?* Harmondsworth: Penguin.
Midgley, Mary. 1996. *Utopias, Dolphins and Computers.* London: Routledge.
Midgley, May. 2013. *Science as Salvation London: Routledge.*
Northcott, Michael S. 2018. Reformed Protestantism and the Origins of Modern Environmentalism. *Philosophia Reformata* 83 (1): 19–33.
Northcott, Michael S. 1996. *The Environment and Christian Ethics.* Cambridge: CUP.
Perkins, David. 2003. *Romanticism and Animal Rights.* Cambridge: CUP.

Pickstock, Catherine. 1998. *After Writing: On the Liturgical Consummation of Philosophy*. Oxford: Blackwell.
Preece, Rod. 2002. *Awe for the Tiger, Love for the Lamb*. London Routledge.
Pollan, Michael. 2002. An Animal's Place. New York Times Magazine, November 10.
Ray, John. 1743. *The Wisdom of God Manifested in the Work of Creation*. London: William Innys.
Rollin, B. E. 2011. Animal Pain: What It Is and Why It Matters. *The Journal of Ethics* 15 (4): 425–437.
Rosenfield, Leonora Cohen. 1968. *From Beast Machine to Man Machine*. New York: Octogon.
Richard D. Ryder. 2016. Are We Addicted to Animal Cruelty? *Sarx* 11.2.2016. http://sarx.org.uk/articles/christianity-and-animals/the-great-addiction/#more-808. Accessed 16.2.2018.
Sampson, Philip J. 2018. *Animal Ethics and the Nonconformist Conscience*. Palgrave Macmillan.
Sampson, Philip J., forthcoming. William Cowper: A Passionate Kindness.
Sandhu, Serina. 2018. The Meat Free Farmer. 'i', March 3.
Sargent, Tony. 1996. *Animal Rights and Wrongs*. London: Hodder.
Santmire, H. Paul. 1985. *The Travail of Nature*. Philadelphia: Forteress Press.
Seidenberg, David Mevorach. 2019. Veganism and Covenantalism. In *Jewish Veganism and Vegetarianism,* ed. Jacob Ari Labendz and Shmuly Yanklowitz, 161–194. New York: Suny Press.
Singer, Peter. 1975. *Animal Liberation*. New York Review.
Singer, Peter. 1999. Reflection. In *The Lives of Animals,* J. M. Coetzee, 85–92. New Jersey: Princeton Univ Press.
Singer, Peter, and Jim Mason. 2006. *Eating*. London: Arrow.
Singer, Peter. 2009. *Animal Liberation*. New York: Ecco.
Singer, Peter. 2011. *Practical Ethics*. Cambridge: CUP.
Stone, Ken. 2018. *Reading the Hebrew Bible with Animal Studies*. Stanford University Press.
Tanzella-Nitti, Giuseppe. 2005. The Two Books Prior to the Scientific Revolution. *Perspectives on Science and Christian Faith* 57: 235–248.
Thomas, Keith. 1983. *Man and the Natural World*. London: Allen Lane.
Toynbee, Arnold. 1974. The Religious Background of the Present Environmental Crisis. In *Ecology and Religion in History, ed.* David and Eileen Spring, 137–150. New York: Harper.
Trapp, John. 1660. *Annotations Upon the Whole Bibl,*.vol. 3 London: R W.
Trapp 1660 vol 1 Genesis 2.20 Names According to their Natures John Trapp (1868:1657) *Commentary on the Old and New Testaments* in 5 vols Dickinson London.
Walker, George. 1641. *God Made Visible in his Works. London:* John Bartlet.

Wallace, Alfred Russel. 1889. *An Exposition of the Theory o f Natural Selection* London: Macmillan.
Warfield, Benjamin B. 1927. *The Works*. Oxford University Press.
Ward, Nathaniel. 1641. *Massachusetts Body of Liberties.*
Wesley, John. 1958:1872. The Imperfection of Human Knowledge. In *Collected Works*, ed. Thomas Jackson, vol. 6, 337–350. Grand Rapids Zondervan.
Westminster Confession. 1647. London: Evan Tyler.
White, Lynn. 1967. The Historical Roots of our Ecologic Crisis. *Science* 155 (3767): 1203–1207.
Wilkie, Rhoda M. 2010. 'Livestock/Deadstock Philadelphia: Temple University Press.
Žižek Slavoj. 1999. *The Ticklish Subject* London: Verso.
Žižek, Slavoj. 2008. *The Sublime Object of Ideology*. London: Verso.

CHAPTER 3

The Choir and the Song

Introduction

The previous chapter introduced the nonconformist understanding of animals as created to praise and manifest God. This chapter explores animal voices as *song*.

Within nonconformist literary culture, the world is not viewed as autonomous *being*, nor are human and other animals seen as self-sufficiently existing. Rather, the world is viewed as "creation", and both human and other animals as dependent on their creator. Moreover, their voices find meaning insofar as they point beyond themselves, longing to manifest wonder and joy. Animals respond to their creator's call to praise, each "according to their kind". I argue that this voice is not primarily "speech" but "song"—it expresses rather than represents. Creation discloses its creator as music discloses the heart of its composer.

The chapter concludes by examining the contemporary implications of this alternative cultural imaginary; one which celebrates all animals as lyric, as meaning, affirming creation as a song sung for another. Creation was, as Jürgen Moltmann (1993, 289) imagined, sung into existence, and that song continues to resound. Creation as song destabilises both anthropocentric hierarchy and zoocentric unity, opening a space for a sustaining and cherishing the displacement of violence.

P. J. Sampson, *Nonconformist Perspectives on Animals and Language*, The Palgrave Macmillan Animal Ethics Series,
https://doi.org/10.1007/978-3-032-12158-5_3

Self-sufficiency and Transcendence

In recent decades, there has been an "animal turn" in a range of academic disciplines as discussion of non-human animals has emerged from the wings. The literature around the ethical interests of animals has grown, together with a recognition of their diverse cultures, although it remains subordinate to the dominant anthropocentrism. Ostensibly this presents a more zoocentric way of speaking of animals as living beings, although as discussed in Chapter 1, it also tacitly relies upon human capacities and attributes. This anthropocentric heritage is not the only feature shared between the dominant discourse and the "animal turn".

As befits modern Western scholarship, both the dominant discourse and the "animal turn" are overwhelmingly secular. Both hang on a shared assumption: the existence of both ourselves and the world as, in some sense, self-sufficient. Emmanuel Levinas (2003, 51) argues that this is a recent development in Western thought, whose historical preoccupation with transcendence and communion with the infinite being has been abandoned, perhaps for the first time. The modern secular sensibility conceives existence as self-sufficient, and it does so with a certainty formerly reserved for creeds. This secular certainty has appropriated man's likeness to God, and transformed it into "sovereignty over existence, the Lordly gaze, the command", attributes which formerly belonged to deity (Horkheimer and Adorno 2002, 6). Self-sufficient and subject to no-one, humans arrogate for themselves the right to speak for all other animals, and, in dominant discourse, to use them for our own ends. The re-orientation of Western thought towards self-sufficiency has largely proceeded without the benefit of interaction with some two millennia of theological reflection upon finitude, and the human condition (Falque 2016).

This confident abandonment of transcendence in favour of self-sufficiency has not gone well. The optimistic expectations of the Enlightenment were shattered by the totalitarianisms of the twentieth century, and more subtle consequences of the doctrine of "progress" are emerging in the twenty-first, not least climate change. Moreover, colonisation by self-sufficiency has desacralised the world, and left a sense of something lost.

Against this background, the nonconformist way of speaking appears like a foreign body, thoroughly subjugated in modernity. It unambiguously affirmed an existence which reaches beyond self-sufficiency, and declared the in-sufficiency of the human condition. I shall argue that precisely its subjugation permits a resistance to anthropocentrism by reimagining the transcendent, leading to insights about the world which are obscured or even invisible to the self-sufficient gaze. By re-appropriating creation as a song sung for another, it provides resources to resist the dominant discourse and explore a genuine "animal turn" which does not hang on humans as its reference point.

Nonconformist Discourse

Dominant and Subsidiary Discourse

For dominant discourse, humans are at the centre, endowed with reason, having governance of other animals which present themselves to us as an inexhaustible supply of fungible commodities. Animals are property, both literally and symbolically. Literally, for their manifest use value. Symbolically, for their use as figures in human culture, icons of civilised triumphs. The bull is not just a bull, he is also a metaphor of strength or potency: the bull of Wall Street whose horns and testicles have been rubbed shiny by countless adoring hands. The pigeon is not just a pigeon, but a metaphor of communication: the rock dove of the RAF's Communication Wing, so necessary for modern warfare.

For the zoocentric subordinate discourse, there is a kinship among all animals, including humans. This ostensibly presents a more humble, less anthropocentric view of humans as one animal among others, albeit one who arrogates for himself the power to both represent and speak for all the others. But it is a "false humility", one which pretends to speak universally, not just for itself. It is unable to reach beyond itself, the "true arrogance" posing as a pretended humility (Žižek 1999, 132–3).

These positions appear radically different: the one tries to control and dominate animals; the other to free them for their own possibilities. Dominant discourse asserts the human-animal binary, while the subordinate discourse struggles to subvert it. However, both return strategically to the place of the human, privileging human power to represent as the origin for thinking about animals, and human power to manifest as the origin of human exploitation of other animals (Derrida 1988, 236).

Both dominant and subordinate discourses begin with self-sufficient existence, the power of representation, and the human-animal binary.

They pose the same anthropocentric question: in this order of things, how does humankind represent its place in the world?

Subjugated Discourse

Before the secularisation of scholarship in the late nineteenth century, a third, now subjugated, nonconformist discourse existed; one which rejected self-sufficiency, autonomous human representation, and the human-animal binary as originary. Rather, the defining feature of the nonconformist gaze is that it "begins, centres and ends on the vision of God in His glory" (Warfield 1981, vol. 5, 358). All creation is entirely dependent, and normed by its chief end: the manifestation and praise of its creator (Westminster Confession 1646, 4.1; 5.1). Moreover, *manifestation* and *praise* go hand in hand. The "glory of God", says John Calvin (1849:1563, 148.3), "is everywhere to be seen, as if they sang his praises with an audible voice". Humans and animals separate out from this underlying unity like rainbow colours from white light. As the Presbyterian leader David Dickson (1959:1653, 527–8) explains: "Beside the common song of praise which all things have as they are creatures, every one of the creatures has its own proper part in the song… every one by themselves, and these also make up a sweet and harmonious melody, to the praise of God"…. "the speechless creatures have their own way of praising God, and men have their own way of praising: speechless creatures praise God in their own kind, when they hold forth... the matter and reason of praising God, and men praise God, when they observe and acknowledge, and proclaim one to another, the reasons for God's praise…".

Living creatures always already possess a "common *song* of praise which all things have as they are creatures", an arche-animality, an "articulating supplement" that both precedes and makes possible the distinctions between animals, and our representations of them (Driscoll and Hoffmann 2018, 11; my italics). Nonconformist discourse's reference beyond itself cuts into and disturbs both the dominant discourse and the "true arrogance" of subordinate discourse posing as a pretended humility.

This theocentrism foregrounded a specific reading of the biblical texts, and led to a distinctive affirmation of existence-as-creation. Sharing creature-hood, all living subjects, humans and other animals, have a unique place in the common creaturely task of manifestation and praise. This is predicated upon mutual engagement, not human representation.

The meaning of existence points beyond creation to the creator, beyond autonomy to inter-dependence, and covenant. Creation speaks of something outside itself, it reveals something other. In this sense, creation has a voice, but it does not simply talk among itself, although it certainly does that. It speaks of another reality, transcendent yet reciprocal.

Covenant and Meaning

Creation does not represent God; rather it *manifests* God in a series of covenants. Two are especially relevant to the present discussion.

Creation Covenant

The creation narratives set out an enacted covenant made with all creation which entails relationship, mutual use and mutual rights (eg Gen. 1, John 1, Col. 1, Heb. 1; Seidenberg 2019, 161–2). The fundamental distinction is not the human-nature binary within creation; but that between the Creator and creation. God the Creator, says Caryl (1652, 207), "betrothed himself in mercies and loving kindness" to all the creatures of earth. Animals, says evangelical polymath, Robert Boyle, are God's creatures equally with us, and he cares for them; sabbath rights are for the animals sake, and exemplify God's "tenderness of Brutes" (Oster 1984). Humans and other animals are alike "creatures", mostly created on the same day. Nowhere does the text of Genesis describe humans as "good". This is a quality accorded to land, sea, vegetation, the heavens, sea-life, birds, and other non-human animals; but nowhere to humans. It seems that the position of humans is to *facilitate* good, rather than *be* good. The "very good" of Gen. 1.31 describes creation as a whole, its harmony, what we might now incline to call its ecology. As Stephen Charnock (1864:1681, vol. 2, 26) put it: all creatures "are so interwoven and inlaid together by the divine workmanship, as to make up one entire beauty in the whole fabric"; they "meet in a common centre, the good and preservation of the universe". God's covenant gives equal footing to all as creatures, and as Seidenberg (2019, 164) argues "ascribes a higher moral standing to animals than has been the norm in modern society".

Nonconformist discourse portrays God as sovereign, and reciprocally engaged with creation; creation as dependent, not self-sufficient. "As God made all things by the word of his command", wrote puritan, John Arrowsmith (1659, 402), "so he *upholds them all in being by the word of his power.* Heaven, earth, sea, man and beast… It is not with God, as with carpenters and shipwrights, who make houses for other men to dwell in, vessels for others to sail in, and therefore after they are made look after them no more; God who made all things for himself, looks to the preservation of all.... The creatures are all as vessels, which if unhooped by withdrawing of Gods manutenency [maintenance], all the liquor that is in them their several vertues, yea their several Beings would run out, and they return to their first nothing". As the Westminster Confession (5.I) summarises: God sustains and preserves his creation "to the praise of the glory of His wisdom, power,justice, goodness, and mercy". The creation was spoken into existence as a peaceable community without either symbolic or manifest violence. Michael Walzer (2012) argues that relations among humans are best rooted in a covenant community's worship of God; the creation covenant is precisely such a community extended to all creatures.

Noahic Covenant

The Fall disrupted this peaceable theocracy, and a new covenant responded to the changed situation. The first covenant was made with all creation; the second, following the flood narrative, with all living creatures, and is oriented to restoration. The text calls it a "perpetual" "covenant of peace", made with Noah and with "every living creature", the latter clause repeated four times for emphasis (Seidenberg 2019, 164). At the "restoration of all things", this covenant will be recapitulated for all creatures with an end to violence, that they may "lie down safely" (Gen. 9; Ezek. 1:28; Isa. 54:9–10; Acts 3.21; Hos 2:18).

The emblem of the Noahic covenant is the rainbow. A bow usually signifies war or hunting, but here, unstrung and hanging in the clouds, it is appropriated as a sign of peace. "It is turned upward; not to shoot arrows down on us, but for us, if we have faith enough to string it, and to make it our glorious bow", extolls Spurgeon (1863, II.2). "It is planted in the clouds", comments Trapp (1660, vol.1, Gen. 9.13), "as if man were shooting at God, and not God at man. This bow with both ends downward, and back to heaven, must needs be an emblem of mercy; for he that shooteth, holdeth the back of the bow from him". As the texts emphasise, this bow is a token of the covenant with all living creatures; no animal is self-sufficient; the meaning of their existence points up, towards God to whom all animals offer praise.

Creation Manifesting Meaning

The world was created for the manifestation and praise of God. Created existence is not autonomous, but neither is it a mere appearance to cover a meaninglessness void. Rather, it *manifests*, pointing beyond itself to the creator as a harmonious ontology of sacred meaning. All God's works, affirms the calvinist and vice-chancellor of Oxford, John Owen (1840:1669, 190; emphasis original), are for "the manifestation and declaration of the holy properties of his nature". The puritan, John Maynard (1688, 11–12, 16), is eloquent on the matter: "If you can once learn to see God in his Creatures" he says, "Thou maist then see him Dayly round about thee: Thou canst not set thy Foot upon the ground, but thou maist remember that thou treadest upon the Lords workmanship, and that thou couldest not stir a Foot, but that he hath made this ground to bear thee, and given thee leave to walk upon it. When thou seest the beautiful face of the Earth, and the Fruits and Plants which it yieldeth thou seest abundant variety of the Lords Creatures, and art taught (if thou wilt learn) to remember him by whom all of them were created. Thou must think thou hearest Him in the noise of strong and mighty winds, in hideous claps of thunder; that thou discernest Him in the Clouds, in the showers of Rain, in the Hail, Snow, Ice, and Hoarfrost; in the Spring, Summer, Harvest, Winter; in all the seasons of the Year; yea in the very Day and Night; for Day unto Day, and Night unto Night, teach knowledge; Time it self, and the continuance of it, are his work". "If in some one of the least Creatures the Lord is to be seen in admirable Workmanship, how much more in the whole creation, especially considering, how it is brought out of nothing? in which respect the least Worm is a witness of his Omnipotency beyond exception; and therefore we must take time for Meditation on the Creatures, that we may set forth the praise and glory of the Creator, and (as the *Psalmist* saith) *May talk of all his wondrous works.* That the Heavens may move us to declare the glory of God, and the *Firmament* may provoke us, and prevail with us to set forth his Handywork".

In particular, this is true of living creatures. The beasts, affirmed Trapp (1868,1657, vol.2, 484) "are good creatures in their own nature and kind, and made to set forth the glory and magnificence of the great God". *Meaning* is the nature of reality answering to the creating word of God. Creation is the ontological disposition that makes meaning possible, not simply its locus.

Many secular authors emphasise that man is made in God's image (Gen 1.26–7) in order to deplore human arrogance. Nonconformist discourse unambiguously sees God's attributes revealed in all creation, not only humans; animals manifesting God in a way that neither humans nor inanimate creation can. The beasts, says Trapp (ibid), are "made to set forth the glory and magnificence of the great God…". From "the angel to the worm that creepeth on the ground", says Caryl (1652, 206), "from all, from the least of these, we may know much of the great God". Puritan, Thomas Manton (1870, vol. 1, 443), agrees: God is seen, not only in man, but in "any gnat or fly"; it is "like the odour of the sweet ointments, or the perfume that he leaveth behind him where he hath been". This "glimpse of the fragile absolute" in worm or gnat can yet bring us to our knees; and God, almighty in gentleness, will not break the "bruised reede" or extinguish the "smoaking flax" (Zizek 2000; Sibbes 1630).

Moreover, we do not need to look far. The glory of God, said Calvin (1845:1559, 1.v.1) is engraved on each of his works so "brightly, distinctly and illustriously" that it is as if "God for the first time was arrayed in visible attire when, in the creation of the world, he displayed those glorious banners, on which, to whatever side we turn, we behold his perfections visibly portrayed". "I can find [God]", says Caryl (1652, 214), "almost without searching, everything I see shows me God". The whole creation, says John Bailey (1689, 9) is "Gods Library, you may read your fill, it was the first Bible that ever God wrote - gracious Souls see God in every thing that he has made, and his glorious perfections in all, and give him the glory of them both in the work of Creation and Providence".

This manifestation of God in creation is, by its very nature, awe-full. Wonder and awe can readily be relegated to a Romantic sensibility but, in nonconformist discourse, they are fundamental in our relationships with animals.

Awe

Awe is a common human response to creation, appropriated by the Romantic movement as the *sublime*, with a particular emphasis upon unsettling and excessive experience. William Wordsworth (1974, vol. 2, 354) spoke of the sublime as of the mind prostrated "before some external agency", attempting to "grasp at something … it is incapable of attaining". His language of posture and touch emphasises bodily response over cognition; the opened mouth, the widened eyes, and changes in the pulse fostered by awe (Fingerhut and Prinz 2018, xxx, 119). This pre-cognitive aspect of awe lends weight to primatologist, Jane Goodall's view that awe is not unique to humans, but shared with other animals, notably chimpanzees (Stone 2018, 142–147).

Many feel that, in a rationalised world, we have lost the sense of something beyond us. Certainly the Romantic movement felt this of Enlightenment rationalism which seemed to neglect everything valued by humans. The Romantic movement's focus on oceans, mountains, especially the Alps, found God imminent in the sacredness of an idealised Nature. Few nonconformists had made the Grand Tour, so most had never seen the Alps. But they had looked up, and the heavens, says the psalmist, declare the glory of God (Ps. 19). For them, awe is our response to the manifestation of God in creation, not sacred imminence. For puritan, Thomas Watson (1965:1692, 10), "the firmament is beautified and pencilled out in blue and azure colours, ... we may see the glory of God blazing in the sun, and twinkling in the stars". But nonconformist discourse especially spoke of awe at animal creation. Rather than the idealised and mystified animals of Romanticism, we find them sympathetically and realistically portrayed as individual creatures (Perkins 2003, 47–8). Awe and wonder here play a special role in the attribution and experience of value and meaning; three components will be important for our discussion: "cognitive perplexity", "perceptual engagement", and a "sense of reverence" (Fingerhut and Prinz 2018; Keltner 2023).

Awe and Pious Delight

For nonconformists, cognitive perplexity in our encounter with creation is not simply the "mind prostrated", but a token of meaning: awe points beyond. Puritan, Richard Baxter (1817:1650, 105), meditating on "those glorious works of creation which our eyes now behold" is rapt: "What wisdom, power, and goodness, are manifested therein! How does the majesty of the Creator shine in this fabric of the world! 'His works are great, sought out of all them that have pleasure therein.' What divine skill in forming the bodies of men or beasts! What excellency in every plant! What beauty in flowers! What variety and usefulness in herbs, plants, fruits, and minerals! What wonders are contained in the earth and its inhabitants; the ocean of waters, with its motions and dimensions; and the constant succession of spring and autumn, of summer and winter!".

Awe gives a glimpse into something at once both smaller than we can imagine; and bigger, deeper, passing all understanding. The German reformer, Martin Luther, struggled to express his wonder, yet he knew where it pointed: the Divine majesty "is so small that it can be substantially present in a grain [of wheat], on a grain, over a grain, through a grain, within and without, and that, although it is a single Majesty, it nevertheless is entirely in each grain separately, no matter how immeasurably numerous these grains may be... And that same Majesty is so large that neither this world nor a thousand worlds can encompass it and say: "Behold, there it is!" (Santmire 1985, 129–30).

Sometimes we don't just struggle for words, but are reduced to silence. Awe becomes a bodily experience, the mind prostrate, but before God. What cannot be captured in speech is present in embodied meaning; we not only see the awe-full beauty, but we want to be perceptually engaged, even united, with it: to enter it and have it enter us.

Awe and Reverence

The glory of God, disclosed in creation denies human self-sufficiency and reaches beyond the experience of our material conditions. For Calvin (1845:1559, 3.X.2) "*pious delight*" points to a *value* beyond use: "Has the Lord adorned flowers with all the beauty which spontaneously presents itself to the eye, and the sweet odour which delights the sense of smell, and shall it be unlawful for us to enjoy that beauty and this odour? ... In short, has he not given many things a value without having any necessary use?".

The sense of reverence characteristic of awe destabilises the anthropocentric perspective of our material conditions, and opens us to new possibilities. This experience may not be safe, for wonder evokes a sense of our own fragile vulnerability which is entirely unknown to the dominant anthropocentric discourse (Fingerhut and Printz 2018, 118). Because "the glory of his power and wisdom is more refulgent in the firmament, it is frequently designated as his palace", says Calvin (1845:1559 1.v.1). Wherever "you turn your eyes, there is no portion of the world, however minute, that does not exhibit at least some sparks of beauty; while it is impossible to contemplate the vast and beautiful fabric as it extends around, without being overwhelmed by the immense weight of glory".

Calvin's sense of being "overwhelmed" is a frequent nonconformist experience. Far from being "monarch" of all they survey[1] in the

[1] The phrase "I am monarch of all I survey" is sometimes taken to exemplify human dominion over animals, a view which seems sustained by the full stanza:

anthropocentric style so often attributed to them, nonconformists were confronted with the fragile vulnerability humans share with all living creatures. "This wonderful work", says Maynard (1688, 14–16, italics original), "viz. the whole World, and all the Creatures in it, should move us to esteem him wonderfull, even beyond astonishment". "And then, if there were not much dulness in our spirits and want of apprehension, there would be (as it were) no spirit in us; the powers of our Souls would even be swallowed up with admiration, and we should with feeling hearts express our astonished thoughts and cry out with the Psalmist, *O Lord our God, how wonderfull is thy Name in all the World! Psal.* 8.1". Maynard felt himself overwhelmed by the "numberless Creatures of divers kinds, their several natures, tempers, qualities, vertues, enclinations some of them great, some lesser, [which] ravish us". He was not alone in using existential language. To truly consider the wonder of the world "would ravish [enrapture] us with the admiration of God", says Charnock (1864:1681,23). "If we truly understood the growth of a grain of wheat", marvels Luther (Santmire 1985, 130), "we would die of wonder". Shorn of its manifesting the transcendent, this wonder and awe was appropriated by the Romantic movement as unsettling and excessive experience, and dismissed by some seventeenth-century anglicans as "Nonsensick Raptures" of puritan preachers (Lane 2011, 101).

Nonconformist awe at animal creation generates a sense of their being-from-elsewhere, not as mythic or fabulous, but as embodied meaning. We find we can no longer speak of them as an empirical object, or even a living subject modelled on human experience, but as uniquely themselves, referring beyond themselves. Animals address us before empirical knowledge is possible; they elude integration into knowledge other than by revelation (Levinas 1969, 49f, 192).

This point of resistance, where nonconformist discourse cuts into and disturbs the secular cosmic order by reference to transcendence, is not abstract, but existential (Žižek 1999, 133). It brings delight to all creation; and, as C. S. Lewis (1993, 94–5) observes, "all enjoyment

"I am monarch of all I survey,
My right there is none to dispute;
From the centre all round to the sea,
I am lord of the fowl and the brute."

In fact, it is an image of solitude which the poet William Cowper (2013:1782, 272) describes as a "horrible place", whose sorrows could be assuaged only "by religion and truth".

spontaneously overflows into praise"; praise not merely expresses but completes the enjoyment; it is its "appointed consummation".

Creation as Praise

Creation not only manifests God. "God, the great Creator", says the Westminster Confession (5.1), upholds all things "to the praise of the glory of his wisdom, power, justice, goodness, and mercy". For Watson (1965:1692, 15) the intimacy between creation and praise is so fundamental as to be etymological: "The Hebrew word *Bara*, to create, and *Barak*, to praise, are little different, because the end of creation is to praise God". This vision of praise is not homogeneous, but according to "kind". All creatures are "bound to praise him [God] in their kind" (Walker 1641, 163). Inanimate and living creation praise differently.

Inanimate Creation

Inanimate creation praises God by the decree that made creatures what they are (Bauckham 2012, 149–50). "By what they are, and their obeying the law of their nature", says Owen (1862:1668, 483), "they give unto God the glory of that wisdom and power whereby they are made, and of that sovereignty whereon they depend".[2] Calvinist, Thomas Adams (1862:1616, vol. 3, 10), writes: "All creatures in their kind bless God. They that want tongues, as the heavens, sun, stars, meteors, orbs, elements, praise him with such obedient testimonies as their insensible natures can afford". Moreover, even with insensate nature, this praise is exuberant. The biblical text is replete with earnest signs of joy: the distant shores rejoice, the earth shouts for joy, the sea, and everything in it resounds, the rivers and trees clap their hands, and the the trees of the forest sing for joy (eg Ps. 97:1, Ps. 98:4, 7–8, Ps. 96:10–13; 100.1–2; Is. 55.12).

Creation's praise is a popular theme of hymnody: Congregationalist, Isaac Watts exhorts, let "Heaven and nature sing"; while William Draper paraphrases St. Francis to implore "All creatures of our God and King, lift up your voice and with us sing… ye lights of evening, find a voice,

[2] Owen is unusual among nonconformists in extending this to non-human animals as a matter of obeying a law, rather than exhibiting a creation norm. See below for the more characteristic nonconformist understanding of an animal's voice.

O praise Him". From Thomas Ken's doxology "Praise God, from whom all blessings flow, Praise him, all creatures here below" to Chris Tomlin's more contemporary declaration that creation reveals God's majesty, everything sings in a unique voice. "Have you never stood by the seaside at night", asked Spurgeon (1885, vol. 44, 1) "and heard the pebbles sing, and the waves chant God's praises?". A vestige of this insight that inanimate creation has a life of its own remains when we speak of the health of a river, or its death by pollution; this marks its place in our cultural imaginary as more than a *thing*, but now floating free of the theocentrism which formerly anchored it (Macfarlane, 2025).

Nonconformist discourse recognises a common creatureliness in inanimate creation which allows rivers and mountains to praise in more than a figurative sense. But it was especially animals, those that have tongues, who are most eloquent.

Animal Creation

By contrast with the passive praise of inanimate creation, animals have a voice; they actively praise. C. S. Lewis (1947, 10) remarks that we "feel the difference between animal voices (including those of the human animal) and all other noises pretty clearly… though it is hard to define". Non-human animals do not speak human languages, but it does not follow that they have no voice. Daniella Rizzo (2024, 60–1), referring to Acts 2, speaks of "animal glossolalia", "transcending the confines of human experience … [to challenge] prevailing anthropocentric perspectives". Thus Adams (1862:1616, 10) says of animals: "they that have tongues, though they want reason, praise him with those natural organs. The birds of the air sing, the beasts of the earth make a noise… the very 'dragons in the deep', … sound out his praise". "Look into the air", says Watson (1965:1692, 10), "the birds, with their chirping music, sing hymns of praise to God. Every beast in its kind glorifies God: 'The beasts of the field shall honour me'". Not only does the praise of rivers and mountains differ from that of living animals, there is also a variety and particularity of praise among the animals themselves. Each animal has its own individual meaning and uniquely manifests God (De Graf 1977, 34).

In the dominant discourse, this pattern of differences is reduced to the "human" and the homogenised "animal", the necessary preconditions for an Enlightenment human-nature binary. The subordinate discourse blurs this boundary in favour of a homogeneous continuity (Driscoll and Hoffmann 2018, 5). Nonconformist discourse rejected both these options, positing a heterogenous multiplicity of creatures, each praising God according to its kind. The animal no longer appears homogenised in "the general singular" of the Enlightenment binary, as fit only for consumption. Animals are diverse, of value individually, not just as a species. You may consider sparrows worthless, says Jesus of Nazareth, but God cares about them individually; in the sight of God they live grievable lives (Matt. 10.29–31; Lk. 12.6; Butler 2020, 24, 56).[3]

For Spurgeon, these different voices join together in a "feast of praise". "The.. entire globe of land and water.. is to be made vocal everywhere with praise. Ye dragons, and all deeps.... Terrible beasts or fishes, whether they roam the earth or swim the seas, are bidden to the feast of praise. Whether they float amid the teeming waves of the tropics, or wend their way among the floes and bergs of polar waters, they ... yield their tribute to the creating Jehovah". It is not only "terrible beasts or fishes" which are "vocal". Echoing Manton's "gnat or fly", for Spurgeon (1885, Ps. 128.10) the croak from the marshes and buzz of a fly also praise God. This calling to praise has a remarkable levelling effect, fatal to the pretensions of anthropocentrism. Manton (1870, vol. 1, 438) again: "... the greatest glory the creature is capable of is to serve the will and set forth the praise of its Creator, for everything that attains not its end is vain. What matter is it whether I be a dog, or a man, a beast, or an angel, if I serve not the end for which I was made?".

In Spurgeon's nineteenth century, few of the animals he mentioned in his sermon had been extensively exploited for human use. As he says, they "pay no service to man; let them the more heartily confess their allegiance to the Lord" (Spurgeon 1885, Ps. 148.7).[4] Animals do not need humans to praise, but when humans recognise the praising voices of other animals,

[3] In recent years, the dominant and subordinate options have also been challenged by a recognition of multiple similarities and differences between and among human and other animals. More directly, Morton has criticised the ontology founding the human-nature binary itself, positing that all reality exists on an equal footing. These developments have obvious similarities with nonconformist ways of speaking. The option of maintaining secularity while seeking a new non-binary ontology is the alternative to challenging alleged secular neutrality. I will not explore it further here. (Plumwood 2012. Derrida 2008. Morton 2016. Harman 2017).

[4] Augustine (1972, XII.4), often cited as a source of anthropocentrism, says much the same thing: "it is the nature of things considered in itself, without regard to our convenience or inconvenience, that gives glory to the Creator".

it can shape human prosody as Spurgeon's rhetoric illustrates. Animals can animate and bring sensuous experience to the "surface of the written page", and even to a Sunday sermon (Driscoll and Hoffmann 2018, 8).

This does not mean that nonconformists like Manton or Spurgeon were unfamiliar with the use of literary tropes. Of course, they knew that to speak of the rivers clapping their hands is personification but, as discussed in Chapter 2, "earnest metaphor" is substantial, it exceeds the figurative, especially the voice of animals.

Tongues like Trumpets

In nonconformist discourse, animals have a tongue, and are vocal in a way that mountains are not. Theirs is not a world of mute experience. They "have a voice", says Caryl (1670, vol. 4, 205), they "have tongues like trumpets to proclaim a providence in ordering the world". So much was clear from Scripture, yet the notion was so far outside the then dominant discourse that many struggled to understand how it could be so.

Some such as Flavel, Caryl, and Owen refer to it as "metaphor" or "metonimy" (sic); but, as discussed in Chapter 2, we should be cautious about reading this through modern spectacles. "Metaphor" could enable things to be said which were simply unsayable in the dominant discourse. The case of fish is striking. Fish "are vocal" with praise, says Spurgeon. Yet, says Caryl (1670, vol. 4, 211–12), fish "are so far from forming up words and sentences, that they cannot make a sound or the rudest noise". Nevertheless, Caryl is clear that "they have a speech though they have no voice", "they say much, though they can say nothing".

Some, grappling with how an animal might "say much, though they can say nothing", refer to "the outward gift of utterance by *gesture*". John Bulwer (1644, 4–7) speaks of "Natures silent motions", which precede words and "doth reach Unto our sense a mouth-excelling speech". He suggests that gesture enabled human-animal communication in Eden. Retained by animals, it is a human deficit, a kind of understanding that Adam "partly lost with his innocency, yet might be repaired in us, by a diligent observation and marking of the outward effects of the inward and secret motions of beasts". This novel reading of the "Schoole of Beastes" anticipates contemporary discussions of the bodily expressivity of language (Viret 1585. Inkpin 2024). If, says Bulwer, "there be some gestures of ours that they [animals] doe not understand, so there are some of theirs which need an Interpreter.. to expound them to us".[5]

Many simply took a practical view. To Wesley (1836, vol. 2, 370), it was clear from both Scripture and creation that, "the actions of animals have an eloquent and a pathetic language. Those that want the help of man have a thousand engaging ways, which, like the voice of God speaking to his heart, command him to preserve and cherish them". For the Calvinist, William Cowper (2013:1785, 121–2, III.326–333), contemplating the cruelties of the hunt, the "silent tears and heart distending sighs" that agonies inspire are "dumb, but yet endued With eloquence".

Provided there is a listener, there is no speech without reply, even if it is silence (Lacan 1977, 40). For Wesley and Cowper, the speech of animals calls for a compassionate listener who will not be silent. As we shall see in the next chapter, there were many who heard, yet whose silent reply was cruelty.

Embodied Praise of Love

Nonconformist discourse portrays animals realistically as *embodied* meaning without being reduced to mere metaphor. A community of flesh which hangs not upon human exceptionalism, but upon a three-fold thread of praise. John King (1864:1594, 310–11), calvinist vice-chancellor of Oxford, exhorts both animals and human infants: "Let both great and small, the infant and dumb beast, sing of thy loving-kindness"; he was, of course, aware that neither infants nor animals are able to offer *rational* or even *verbal* praise, yet both can offer *authentic* praise. Contemporary scholarship is rediscovering the reality hidden behind the anthropocentric assumption that animals can only praise figuratively. Capacities once thought unique to humans are being recognised in other animals, including language and "expressions of spirituality" (Stone 2018, 142f). And just as we interpret and elaborate one-another's worship as humans, so interpretation of praise by other animals is both possible and desirable. Language without meaning signifies nothing, and animal praise

[5] John Bulwer opposed nonconformity during the civil war, but had been tutored by the Puritan inclined John Walsall, and was known for his Baconian views; his writing shows a strong dependence upon biblical rather than traditional sources, and is largely compatible with nonconformist discourse when speaking of creation.

has meaning, as does glossolalia, which precedes our ability to express it in human words.

Human praise to God is a surplus to what we do as autonomous material beings. We can build, speak, sing, and dance without praise; but we feel the lack. If we relinquish this surplus, we relinquish the comeliness of meaning: "it is good to sing praises unto our God; for it is pleasant; and praise is comely" (Ps. 147.1). Just as taste is surplus to the nourishment which makes food more than fuel, so praise is the central supplement making embodied existence substantial. It is what points everything beyond itself as the simply material. But not only humans can enjoy the taste surplus to nourishment. "A beast tasteth as much of the sensitive sweetness of his food and ease as you do" says Baxter (1670, p184-5); for Maynard (1688, 5), they do "that which is pleasing to them… refusing things distastfull". Similarly, not only humans relish praise. Praise emphasises that all animals are creatures, not the commodities of the dominant discourse; it restores agency to material existence by giving animals a voice which demands to be taken seriously, and to be interpreted.

Singing Praise

In nonconformist discourse, language is fundamentally "doxological", an expression of praise, and a dwelling place for glory. Yet, creation is called upon to *sing* praise, not simply speak it. We have already seen examples by King, Watson, Adams, Spurgeon, and Watts, among others, who drew heavily upon biblical pericopes which exemplify and commend the close relationship between song and praise (eg Ps. 96; Judges 5.3:, Job 36.24; Ps 28.7; Is. 42.10; Neh. 12.46; etc.). Pickstock (1998, 7) points to a Christian tradition emphasising harmonious musical relationships within and beyond creation; music, she says, is "the science that most leads towards theology". The dominant discourse uses the language of property to describe animals, and the subordinate discourse uses that of biology to characterise living beings; in both cases animals are understood as *autonomous* beings. The now subjugated discourse of nonconformists speaks of animals as praise, sung for another, and therefore not autonomous. Song has words but also music; range, timbre, key, and tonal centre. In a good song, the music is more than incidental. It expresses what we can't quite put into words, what we turn to in moments of hope and longing, of celebration and lament. Such nonconformist 'song' was subsequently appropriated by the Romantic movement and transformed into human spiritual renewal. The deep sea, says Byron, (1888, 4.178; 258) has "music in its roar"; an "eternal harmony" which binds "all things with beauty". But sufficient of the nonconformist's biblical heritage persists to enable Derrida (2008, 53, 63) to meditate on how to have himself hear in a language other than the inarticulate cries which humans attribute to other animals; an "unheard-of music", he says, with which to denounce anthropocentrism.

If animals are to sing, they must have a musical voice. We recognise bird song, whale song, and cicadas singing as though "their music exceeds any possible functional purpose" (Pickstock 2007, 53); "the praises of God in the sweet song of feathered minstrels", as Spurgeon (1885, Ps. 148.10) puts it. The dawn chorus, says Horatius Bonar (1875, 41), minister of the Free Church of Scotland, turns "all the air to music". In a text hocketing between biblical pericopes and his own voice, he celebrates the songs of birds: "Thus the air is vocal. It has a hallelujah of its own. The 'flying fowl' praise him; whether it be 'the stork that knoweth her appointed time' (Jer. 8:7), or 'the sparrow alone upon the housetop' (Ps. 102:7), or 'the raven of the valley' (Pr. 30:17), or the eagle 'stirring up her nest, and fluttering over her young' (Deut. 32:11), or the turtle making its voice to be heard in the land (So. 2:12), or the dove winging its way to the wilderness (Ps. 102:6). This is creation's harp (truer and sweeter than Memnon's) which each sunrise awakens".[6]

We can hear this song, although it is not primarily sung for us but to God. There are many songs we cannot hear, simply because our hearing is not accommodated to them; and yet others which we choose not to hear. The unheard music of mice is largely ultrasonic; male mice sing complex songs to attract females, just like songbirds; only recently have we become aware of this (Driscoll 2017). Songs unheard because we do not wish to hear them include the "croaking from the marshes" and the buzz of a fly, yet are, as Spurgeon makes clear, all *songs* of praise. The failure is in our hearing, not their singing.

Song constrains the symbolic violence tied to human representation, both because all creation sings, not only humans, and because song primarily expresses rather than represents; if we could find some other

[6] "Memnon's harp" refers to the sound made by an ancient Egyptian statue at sunrise, believed to be the voice of Memnon greeting his mother.

way of expression, it would not be authentic song (Pickstock 2007, 41). As awe beyond words finds bodily expression, so song ties praise to the body. The birds open their *beaks* to sing, the lamb *leaps* for joy. But even as it constrains symbolic violence, this bodily attachment also raises the possibility for silencing song, for opening a space for the cruelty of the suffering body, as we shall see in Chapter 5 (Falque 2018).

Creation's Hallelujah

A. A. Hodge (1890, 76), presbyterian principal of Princeton College, contrasts human musicians, who makes music from instruments designed and made by others, with God who is both instrument-maker and composer. Speaking of the Book of Scripture, Hodge describes the God who builds and tunes his instrument for "the very qualities he needs", fashions the orchestra and composes the score, bringing "the very music in its fulness which his soul has designed from the first". So, also, for the Book of Creation. "The whole world is a musicall instrument; every string is so tuned, as to be fitted for resounding the great Creators prayse". So wrote puritan, John Arrowsmith (1645, 23). God is at once the instrument-maker and composer, the craftsman and librettist, in both the Book of Creation and that of Scripture.

This nonconformist discourse is entirely foreign to modern readers. Songs in the modern world have always been implicitly human, invoking a human presence and stable point of origin; bird "song" can only be figurative (Driscoll 2017). This is not so of nonconformist literature. They inhabited a cultural imaginary which found its stable point of origin outside the humanist frame, in the transcendent presence of God. Indeed, its celebration of divine sovereignty sometimes appeared anti-humanist to its critics, exemplified in the eighteenth century debate about calvinism's alleged denial of human free-will.

Creation discloses its creator as the music discloses the heart of its composer. "God must be glorified in all creatures, because hee is the maker of all creatures", says Calvin (1845:1559, 1.v.5). "The End, why all is made, is for the glory of the Maker" echoes Adams (1862:1629, vol. 3, 120). God must be glorified in creation's song because God is the composer of song. Moreover, God did not compose, and then leave the orchestra to it. Rather, He sustains the harmony, not by "producing a kind of general motion in the machine of the globe as well as in each of its parts" but by a "special providence sustaining, cherishing, superintending, all the things which he has made, to the very minutest, even to a sparrow", and presumably to the sparrow's song (Calvin 1845:1559, 1.xvi.1).

Only singers are found "sustaining [and] cherishing" song. It would be strange if the composer of music were not musical, if the composer of songs did not sing. It is, after all, "God my Maker, who giveth songs in the night" (Job 35.10). God "is the great Composer of songs and Teacher of music, he it is who can teach thee how to sing", says Spurgeon. This nonconformist trope opens a space for reality as song, sung into existence.[7] C. S. Lewis (2009, 116–127), who was not in the nonconformist tradition but who was familiar with the literature, imagined just such a singing in his account of the creation of the fictional world of Narnia.

[In an utterly dark void, a] "voice had begun to sing. It was … hard to decide from what direction it was coming. Sometimes it seemed to come from all directions at once. Sometimes … out of the earth beneath.. Its lower notes were deep enough to be the voice of the earth herself. … [The] voice was suddenly joined by other voices; more voices than you could possibly count. They were in harmony with it, but far higher up the scale: cold, tingling, silvery voices. The … blackness overhead, all at once, was blazing with stars. … One moment there had been nothing but darkness; next moment a thousand, thousand points of light leaped out—single stars, constellations, and planets … The new stars and the new voices began at exactly the same time. … it was the stars themselves who were singing, and … the First Voice, the deep one, … had made them appear and made them sing.… The Voice rose and rose, till all the air was shaking with it. And just as it swelled to the mightiest and most glorious sound it had yet produced, the sun arose.… When a line of dark firs sprang up on a ridge about a hundred yards away …they were connected with a series of deep, prolonged notes … sung a second before. … [A] rapid series of lighter notes [saw] primroses suddenly appearing in every direction".

If reality were sung into existence as Lewis imagines, it retains this quality of song; its notes continue to resound, echoing through wood and field, cattle and free-living creatures.

[7] According to Bruce Chatwin (2012, 2), aboriginal creation myths also speak of singing the world into existence.

Jürgen Moltmann (1993, 289, 183; also 245) imagines a more theological expression of creating song: "In the quickening breath [of the Spirit of God], and through the form giving word [of God], the Creator sings out his creatures in the sounds and rhythms in which he has his joy and his good pleasure". The creation also will share this joy: the "bliss of the new creation finds expression in the eternal song of praise". Language, as Pickstock (2007, 58) argues, is fundamentally "doxological", an expression of praise. To neglect this is to isolate the imminent from the transcendent; the "horizontal" life events from "vertical", liturgical adoration of God. The ontological, she says, "is self-exceeded as doxological, as praise of the divine, because it is inversely exceeded by the divine descent into time", "the divine reaching down and restoring, in Christ, of a true temporal ontological sequence as a life of worship".

The Whole Story

I have argued that the now subjugated discourse of nonconformists spoke of animals as joyfully singing God's praise in a way that inanimate creation does not; an animal glossolalia.

I have mostly spoken of individual animals or animal "kinds". I have in passing noted that animals sing together, in Dickson's phrase cited above, making "a sweet and harmonious melody, to the praise of God". However, I have given little detail about this relationship between creatures, nor about the place of humans in this song. I have also not addressed the obvious objection that realities such as predation seem out of tune. Darwin (1860) famously cited the example of the Ichneumon wasp which eats its host from the inside as evidence against a song of "design and beneficence" in creation. The next chapter addresses creatures singing together, and the harmony they produce; the following two chapters treat respectively the place of humans, and the dissonance of a suffering creation.

References

Adams, Thomas 1862:1616. The Taming of the Tongue. In *The Works of Thomas Adams*, vol. 3, 10–22. Edinburgh: James Nichol.

Adams, Thomas 1862:1629. Meditation upon Some Part of the Creed. In *The Works of Thomas Adams*, vol. 3, 85–264. Edinburgh: James Nichol.

Arrowsmith, John. 1645. *Englands Ebenezer.* London: Samuel Man.

Arrowsmith, John. 1659. *Armilla Catechetica. A Chain of Principles.* Cambridge: John Field.
Augustine. 1972. *The City of God.* Penguin.
Bailey, John. 1689. *Man's Chief End to Glorifie God.* Boston: Samuel Green.
Bauckham, Richard. 2012. *Living with Other Creatures.* Milton Keynes: Paternoster.
Baxter, Richard. 1817:1650. *The Saints' Everlasting Rest.* Glasgow: Blackie.
Baxter, Richard. 1670. *The Life of Faith in three parts...* London: Nevil Simmons.
Bonar, Horatius. 1875. *Earth's Morning.* New York: Robert Carter.
Boyle, Robert in Malcolm R Oster (1984) The 'Beame of Diuinity'. British Journal for the History of Science, 22: 151–180.
Bulwer, John. 1644. *Chirologia.* London: Tho. Harper.
Butler, Judith. 2020. *The Force of Non-violence.* London: Verso.
Calvin, John. 1845:1559. *The Institutes of the Christian Religion.* Macdonald.
Calvin, John. 1849:1563. Commentary on the Psalms. Edinburgh: CTS.
Caryl, Joseph. 1652. *An Exposition upon the Book of Job (Chapters 11–14).* London: J. Macock.
Charnock, Stephen. 1864:1681. Discourse on God's Wisdom. In *Works,* vol 2. Edinburgh: James Nichol.
Chatwin, Bruce. 2012. *The Songlines.* London: Random House.
Cowper, William. 2013:1782 & 1785. *The Task and Select Other Poems.* ed. James Sambrook. London: Routledge.
Darwin, Charles. 1860. Darwin Correspondence Project, "Letter no. 2814," Accessed on 26 April 2025, https://darwinproject.ac.uk/letter/?docId=letters/DCP-LETT-2814.xml.
De Graf, S.G. 1977. *Promise and Deliverance.* New Jersey: Presbyterian and Reformed.
Derrida, Jacques. 1988. *Limited Inc.* Evanston: Northwestern University Press.
Derrida, Jacques. 2008. *The Animal that Therefore I Am,* ed. Marie-Louis Mallet. trans. David Wills. New York: Fordham University Press.
Dickson, David. 1959: 1653. A Commentary on the Psalms. London: Banner of Truth.
Driscoll, Kári. 2017. An Unheard, Inhuman Music. *Humanities,* 6 (2):art. no. 26.
Driscoll, K., and E. Hoffmann. 2018. Introduction. In What Is Zoopoetics?, ed. K. Driscoll, E. Hoffmann, 1–13. Palgrave Macmillan.
Falque, Emmanuel. 2016. *Crossing the Rubicon.* New York: Fordham University Press.
Falque, Emmanuel. 2018. *The Guide to Gethsemane.* New York: Fordham University Press.

Fingerhut, Joerg, and Jesse Prinz. 2018. Wonder, appreciation, and the Value of Art. In *The Arts and The Brain*, ed. Julia F. Christensen and Antoni Gomila, 107–128. London: Academic Press.
Habermas. Jütgen. 2002. Religion and Rationality. Cambridge: Polity.
Harman, Graham. 2017. *Object Oriented Ontology*. London: Pelican.
Hodge, A. A. 1890. *Evangelical Theology*. London: T Nelson.
Horkheimer, Max and Theodor Adorno. 2002. *Dialectic of Enlightenment*. Stanford University Press.
Inkpin, A. 2024. Merleau-Ponty, Taylor, and the Expressiveness of Language. *Southern Journal of Philosophy* 62: 358–373.
Keltner, Dacher. 2023. *Awe*. Harmondsworth: Panguin.
King, John. 1864:1594 Lectures upon Jonah. Edinburgh: James Nichol.
Lacan, Jacques. 1977. *Écrits*. London: Tavistock.
Lambert, Gregg. 2016. *Return Statements: The Return of Religion in Contemporary Philosophy*. Edinburgh: Edinburgh University Press.
Lane, Belden C. 2011. *Ravished by Beauty*. OUP.
Levinas, Emmanuel. 1969. *Totality and Infinity*. Pittsburgh: Duquesne University Press.
Levinas, Emmanuel. 2003. *On Escape*. Stanford University Press.
Lewis, C. S. 1947. *Perelandra* London:Macmillan 1947.
Lewis, C. S. 1993. *Reflections on the Psalms*. London: Geoffrey Bles.
Macfarlane, Robert. 2025. *Is a River Alive?* Penguin.
Manton, Thomas. 1870. *The Complete Works* London: James Nisbet
Maynard, John. 1688. *The Beauty and Order of the Creation*... London: William Gearing.
Moltmann, Jürgen. 1993. *The Way of Jesus Christ: Christology in Messianic Dimensions*. trans. Margaret Kohl. Minneapolis: Fortress Press
Morton, Timothy. 2016. *Dark Ecology*. New York: Columbia University Press.
Oster, Malcolm R. 1984. The 'Beame of Diuinity.' *Br. J. for the History of Science* 22: 151–180.
Owen, John. 1862:1668. A Practical Exposition upon Psalm 130. In *Works*, vol. 6, 324–647. Edinburgh: T & T Clark.
Perkins, David. 2003. *Romanticism and Animal Rights*. Cambridge: CUP.
Pickstock, Catherine. 1977. Music, Soul, City and Cosmos after Augustine. In *Radical Orthodoxy*, ed. John Milbank, Catherine Pickstock, and Graham Ward, 243–277. London: Routledge.
Pickstock, Catherine. 1998. The Musical Imperative. *Angelaki* 3 (2): 7–19.
Pickstock, Catherine. 2007. God and Meaning in Music *Sacred Music*, 134 (4): 40–62.
Plumwood, Val. 2012. *The Eye of the Crocodile*. ANU Press.
Rizzo, Daniella. 2024. Animal Glossolalia: A Pneumatological Framework for Animal Theology. *Pneuma* 46: 60–79.

Paul, Santmire H. 1985. *The Travail of Nature*. Philadelphia: Fortress Press.
Seidenberg, David Mevorach. 2019. Veganism and Covenantalism. In *Jewish Veganism and Vegetarianism,* ed. Jacob Ari Labendz and Shmuly Yanklowitz, 161–194. New York: Suny Press.
Severan, A. 2021. *Metamodernism and the Return of Transcendence.* Palimpsest Press.
Sherwood, Yvonne, and Kevin Hart, eds. 2004. *Derrida and Religion: Other Testaments*. London: Routledge.
Sibbes, Richard. 1630. *The Bruised Reede and Smoaking Flax. London:* R Dawlman.
Spurgeon, C.H. 1863. The Rainbow. *Met. Tab. Pulpit,* vol. 9, June 28.
Spurgeon, C.H. 1885. *The Treasury of David.* London: Marshall.
Spurgeon, C.H. 1885. Songs in the Night. *Metropolitan Tabernacle Pulpit* vol. 44, February 27.
Stone, Ken. 2018. *Reading the Hebrew Bible*. Stanford University Press.
Thomas, Keith. 1983. *Man and the Natural World.* London: Allen Lane.
Trapp, John. 1660 *Annotations Upon the Whole Bible.* London: R W.
Trapp, John. 1868:1657. *Commentary on the Old and New Testaments.* London: Dickinson.
Viret, Pierre. 1585. *The Schoole of Beastes.* London: Robert Wal-de-Grave.
Walker, George. 1641 *God Made Visible in his Works.* London: John Bartlet.
Walzer, Michael. 2012. *In God's Shadow*. Yale University Press.
Warfield, B. B. 1981. *The Works of Benjamin B. Warfield.* Grand Rapids: Baker.
Watson, Thomas. 1965:1692. *A Body of Divinity.* Edinburgh: Banner of Truth.
Wesley, John. 1836. *A Compendium of Natural Philosophy.* London: Thomas Tegg.
Wordsworth, William. 1974. The Sublime and the Beautiful. In *Prose Works,* ed. W.J.B. Owen & J. Worthington Smyser, 349–360. Oxford: Clarendon Press
Žižek, Slavoj. 1999. *The Ticklish Subject*. London: Verso.
Žižek, Slavoj. 2000. *The Fragile Absolute*. London: Verso.

CHAPTER 4

Singing Together

Introduction

In the previous chapter, I described a nonconformist ontology of meaning, in which creatures sing praises to their creator even as they manifest His attributes. Various images suggested a harmony of individual voices: a choir, a feast of praise, an orchestra, and the tuned strings of a musical instrument. In this chapter, I will be concerned primarily with the harmonies of creatures; and with the meaning of their singing together. This opens a space to talk about animals, both as having individual value, and as finding a meaning in relation to others.

The chapter falls into two parts. It begins with the dominant Western understanding of relationships between human beings and animals, exemplified by the hierarchy of the Great Chain of Being. This paves the way for the second part which contrasts hierarchy with the idea of harmony foregrounded in nonconformity. While a hierarchical understanding of relationships between animals and human beings foreshadowed the dualistic instrumentalism of the Enlightenment, the notion of harmony emphasises mutual dependence within covenants established by God. All creatures, including human beings, are inter-dependent within a common calling to praise God, foreshadowing the "modern idea of the balance of nature" (Thomas 1983, 278). Creation "singing together", weaving

P. J. Sampson, *Nonconformist Perspectives on Animals and Language*, The Palgrave Macmillan Animal Ethics Series,
https://doi.org/10.1007/978-3-032-12158-5_4

diversity into harmony, gifted the nonconformists a point of resistance to the dominant and exploitative discourse of property and instrumental use.

Chain of Being

From the early medieval period to the late eighteenth century, the dominant discourse for relationships between created beings was the Great Chain of Being, accepted by "most educated men... without question". Its links form a ladder of perfection, and run "in hierarchical order from the meagerest kind of existents, which barely escape nonexistence, through 'every possible' grade up to ... the highest possible kind of creature" (Lovejoy 2001, 59, 55). Every creature exists for the sake of the next in the Chain and, in that restricted sense, is dependent; self-sufficiency is an attribute of deity as there is no next, higher link.

In this discourse, the universe is constituted organically as a hierarchy of creatures. The relationship between creatures is governed by orderliness. For a coherent whole, each creature must be in its proper place in the Chain, exercising legitimate power over the less perfect. This is echoed in the social sphere, in which each should keep to their allotted place, exercising power over those lower in the hierarchy. The hymn "The rich man in his castle, The poor man at his gate, God made them, high or lowly, And ordered their estate" is entirely colonised by Chain of Being discourse (Alexander 1848). Thus the order of society is maintained and, as microcosm, it recapitulates the macrocosm of the order of the universe. Conversely, evils arise when individuals do not keep to their proper place; this threatens the coherence of both the macrocosm and the social order. Justice re-establishes the cosmic balance.[1]

Both Lord and Bondsman

However, there is point of instability here. On the one hand, the fact that, say, plants were created for the sake of herbivores indicates the greater perfection in the Chain of Being of herbivores over plants. But it could also be said that herbivores are dependent upon plants to live, challenging their relative perfection over plants. Moreover, since uncreated deity is self-sufficient, deity cannot be dependent upon anything, so no creature

[1] For a discussion of hierarchy and similitude in the Great Chain, see Preece and Fraser (2000, 248–250).

can have been created for the sake of deity. Yet every creature exists for the sake of the next in the Chain, and deity is the culmination of the Chain. It seems that creatures (and worse, deity) are both lords and bondsmen, both more perfect and dependent (Hegel 2019, 91–97).

Leaving aside the question whether the gods are within the Chain, this dialectical instability deepened during the Renaissance with the resurgence of classical literature portraying the gods precisely as dependent upon humans to provide for what they lacked: food and other services for their feasts. Many Christian authors responded by rejecting the notion that creation existed "for the sake of" God, as though He lacked something; God is self-sufficient. Rather, creation exists for the sake of man only, and not for God's necessary use. We frequently find remnants of this dialectic within nonconformist discourse, as well as attempts to be free from it.

An Unfolding Story

Chain of Being discourse has multiple origins, particularly in the work of Aristotle which was adopted by the foremost medieval theologian, Thomas Aquinas (hereafter, Thomas) who explicitly read the biblical texts through its lens.

The Pre-modern Synthesis

Thomas appropriated and transformed *dominion* and *image of God* (Gen. 1.26–8), by reading them within Chain discourse, in which the ontologically imperfect is *for the sake of*, and hence for the *use of*, the more perfect. Man's dominion over the animals became the rightful exercise of power over them; he may, for example, hunt and kill them. As Thomas writes:

In "the order observed by nature... we perceive a certain order of procession of the perfect from the imperfect ... [So] also is there order in the use of natural things; thus the imperfect are for the use of the perfect; as the plants make use of the earth for their nourishment, and animals make use of plants, and man makes use of both plants and animals. Therefore it is in keeping with the order of nature, that man should be master over animals. Hence the Philosopher [i.e. Aristotle] says (*Polit.* i, 5) that the hunting of wild animals is just and natural, because man thereby exercises a natural right".

> "Secondly, this is proved by the order of Divine Providence which always governs inferior things by the superior. Wherefore, as man, being made to the image of God, is above other animals, these are rightly subject to his government".

The ethics of "use" are predicated on ontology, the "order of procession of the perfect from the imperfect" being. Where Thomas finds in the biblical text a restriction on this "natural right", as in the case of the prohibition of human cruelty towards animals, this cannot be *for the sake of* the animal, as animals are ontologically inferior to humans:

"Wherever in Holy Scripture there are found prohibitions of cruelty to dumb animals, as in the prohibition of killing the mother-bird with the young (Deut. xxii, 6, 7), the object of such prohibition is either to turn man's mind away from practising cruelty on his fellow-men, lest from practising cruelties on dumb animals one should go on further to do the like to men, or because harm done to animals turns to the temporal loss of man, either of the author of the harm or of some other; or for some ulterior meaning, as the Apostle (1 Cor. ix, 9) expounds the precept of not muzzling the treading ox" (Thomas. 1924, III.II.cxii; 1947 II.I, Q102).

For Thomas, God rules animals for the sake of "rational creatures", namely men; nonconformist discourse emphasised that God cares for animals for their own sake. Yet, for Thomas, "for the sake of" is not yet the proprietorial language of *instrumentaluse* reached in the eighteenth-century Enlightenment, not least because there was no single understanding of ownership such as was established in early capitalism. However, the language of being *for the sake of*, and for the *use* of, readily lent itself to a subsequent discourse of instrumental use. As we shall see, it also introduced a tension in seventeenth-century nonconformist discourse.

Yet, even as Thomas appropriates the first chapter of Genesis into a Chain of Being, the biblical text constantly resists this appropriation, and threatens to destabilise his synthesis. We have seen this in Thomas' gloss on the biblical prohibition of cruelty towards animals; the opposition to cruelty in the biblical texts is an irritating burr within Chain discourse, and requires special pleading to accommodate it.

But other biblical texts are harder to accommodate, and risk traumatically upsetting the microcosm of social harmony in which everybody occupies their rightful place. According to Luke's gospel, not only is the son of a woman of "low estate" to become king, but the "mighty" will be "put down" "from their seats", and those of "low degree" exalted (Luke 1.32–55; 16.19–31). This scandalous rejection of the ethic of "everything in its right place" steps beyond the way things are. It introduced an individual freedom with respect to community of birth unknown to Greek society, just as care towards animals is foreign to a hierarchy of perfection. Nonconformist discourse embraces such fluidity.

Renaissance humanism introduced further tensions into the Great Chain of Being.

The Renaissance

Firstly, the Renaissance study of a wide range of classical literature disclosed accounts of the gods creating humans for their own use. To speak of the Christian God in this way risked destabilising His self-sufficiency as though he lacked something, requiring human services like the pagan gods. It was preferable to say that creation was made solely for the sake of man.

Secondly, Renaissance humanism primarily concerned itself with mankind rather than with theology. Unlike other animals who are governed by their place in the Chain, humans acquired a new freedom. For example, Pico della Mirandola (1956, 9, 7) refers to man as a "chameleon", given no proper place in the Chain at creation, but able to select his own, to "have and possess through your own judgement and decision". All other creatures have their allotted nature, but humans may "by your own freewill" "trace for yourself the lineaments of your own nature": "placed at the very centre of the world, so that from that vantage point, you may with greater ease glance round about you on all the world contains". Where animals, like the rest of Nature, are governed by their place in the Chain, man is now unconditionally free. Combined with the trope that creation was made for the sake of man (and not God), the ground for an anthropocentric self-sufficiency has been laid, and for a dualism between "governed" Nature and "free" Man.

An Essay on Man

By the early eighteenth century, a fluid and somewhat unstable assemblage of ways of speaking about animals had developed. Motifs from the Great Chain of Being, the biblical texts, and from an emerging humanism, interwove in complex patterns of appropriation and conflict. Poet and essayist, Alexander Pope, reflects this (Pope 1912:1734, I.VIII.5–12; III.I.22–23). With Thomas, Pope celebrates the:

> "Vast chain of being! which from God began,
> Natures aethereal, human, angel, man,
> Beast, bird, fish, insect, what no eye can see...".

A Chain in which all creatures are governed by their place, and which guarantees the coherence of the whole: "One all-extending, all-preserving soul, Connects each being, greatest with the least". If the Chain is broken, it will leave a void, "one step broken, the great scale's destroy'd". The consequence would be a general dissolution of the cosmic order; the world would cease to be coherent.

Yet, Pope also embraces a Renaissance humanism far removed from Thomas, and anticipating the flowering of the Enlightenment:

> "Know then thyself, presume not God to scan;
> The proper study of Mankind is Man".

For Pope, man finds fulfilment in accepting his place, not so much in a Chain of Being, but in an ordered, *rational* world. This humanism commended him to a variety of Enlightenment figures, including Voltaire, Rousseau, and Kant (Solomon 1993, 1; Damrosch 2005, 156).

A committed Catholic, Pope was familiar with the biblical text, and resisted following Thomas in accommodating it to Aristotle. As we have seen, Thomas considers that any "prohibitions of cruelty to dumb animals" found in Scripture are to prevent loss to the man, not pain to the animals. By contrast, for Pope the biblical texts prohibit cruelty towards animals because cruelty is wrong. Notably declining to appropriate *dominion* to Aristotle, he wrote in The Guardian in 1713, that "mankind are no less in proportion, accountable for the ill use of their dominion over creatures of the lower rank of beings, than for the exercise of tyranny over their own species. The more entirely the inferior creation is submitted to our power, the more answerable we should seem for our mismanagement of it" (Pope 1713, 61). Talk of "the lower rank of beings" and our power over "the inferior creation" uses the Chain language; yet Pope here resists the appropriation of dominion in a way reminiscent of the nonconformist discourse of previous centuries. For example, John Calvin (1583, 755) required accountability for our dominion, preaching that his farming congregation should accustom themselves "to bee kinde hearted, and to doe no deede of crueltie, as that even the little birdes may feele it...[for a] young bird hath no tong [sic] to crave reason at our hands".

Similarly, where Thomas follows Aristotle in finding hunting a "natural right", Pope (1720, 12) rejects this in favour of compassion:

> See! from the brake the whirring pheasant springs,
> And mounts exulting on triumphant wings.
> Short is his joy; he feels the fiery wound,
> Flutters in blood, and panting beats the ground....
> Oft, as the mounting larks their notes prepare,
> They fall, and leave their little lives in the air.

Free of Chain language, this is more reminiscent of William Cowper's (1967:1785, book VI, 389–391) complaint against hunters, "why bird and beast Should suffer torture, and the streams be dyed With blood of their inhabitants impaled", than it is of Thomas.

Moreover, Pope's (1912:1734, III.IV.6–18) discussion of animal slaughter robustly prioritises the biblical texts over both Chain discourse and the emerging humanism. He emphasises, not lack of coherent order, but a contrast between a prelapsarian world of creation's praise, and a post-diluvian world which groans under sin, intertexting Genesis 1 and Romans 8.26:

> "Man walk'd with beast, joint tenant of the shade.
> The same his table, and the same his bed;
> No murder cloth'd him, and no murder fed.
> In the same temple, the resounding wood,
> All vocal beings hymn'd their equal God:
> The shrine with gore unstain'd, with gold undrest,
> Unbrib'd, unbloody, stood the blameless priest:
> Heav'n's attribute was universal care,
> And man's prerogative, to rule, but spare.
> Ah! how unlike the man of times to come!
> Of half that live, the butcher, and the tomb;
> Who, foe to nature, hears the gen'ral groan,
> Murders their species, and betrays his own".

Enlightenment Instrumentalism

While both the determinism of Nature, and the humanism and self-sufficiency of man were adumbrated in the Renaissance, the Enlightenment saw them step into the light. Man, come of age, emerged into a new freedom "from his self-imposed immaturity" (Kant 2009, 1). Animals, part of nature, remained subject to law. The complex and unstable relationship between creatures described by Pope collapsed into a dualism: the rational determinism, even mechanism, of Nature; and the freedom and humanism of Man. The Enlightenment built the discursive twin pillars of Western thought: Reason as the royal road to scientific knowledge; and Man as the "titular lord of nature", "his vocation to be the ultimate end of nature" (Kant 2000, 298, para. 83). Ironically, Man had a specific place restored to him: that of the rational animal, the sovereign end of Nature, free to exercise unconstrained power over it. So "far as animals are concerned", says Kant, "we have no direct duties. Animals are not self-conscious and are there merely as a means to an end. That end is man" (Kant 1930, 239). Animals exist for man, but man exists as an end in itself, he has become self-sufficient, the centre of coherence and meaning.

Enlightenment discourse saw the emergence of a genuine instrumentalism. Animals, says Kant (1930, 239–41) bluntly, "are regarded as man's instruments". Similarly, ethics also revolved around man: a man is under no ethical obligation to avoid cruelty to animals, save that he harms "the kindly and humane qualities in himself".

In some ways this was even more anthropocentrically radical than the ontological inferiority of animals in the Great Chain of Being; humans had become self-sufficient, like gods as the source of values; animals are mere instruments for his use. This dominant discourse of animals reached its present form when Enlightenment instrumentalism was integrated into capitalist practices of property ownership. Fragments from the Biblical texts such as *dominion* were appropriated and transformed as they took up their place within this new anthropocentrism. To say creation exists "for the sake of man" now acquired a fully anthropocentric and instrumental meaning. Evils arise, not from disorderliness in the Chain, but from men not behaving according to the light of Reason. None of these various transformations provided much comfort for animals suffering at human hands.

Hierarchy and Freedom

Running through these changing ways of talking about the relationships between creatures, from Chain discourse to that of the Enlightenment, were identifiable threads, albeit nuanced in different ways:

1. The relationship among creatures is hierarchical, placing humans above animals whether in perfection, freedom or rationality.
2. Humans have a freedom not seen in animals, whether on account of dominion or Reason. Animals, by contrast, are subject to a fixed place in the Chain of Being, or are part of nature; in either case, they lack the freedom characteristic of humans. This dualism juxtaposed Humans with Nature (including animals), freedom with determinism.
3. Humans have unrestricted use of animals, either by virtue of their respective places in the chain, or instrumentally.
4. Evils arise when the coherence of creation becomes disordered either by disruption of the Chain or by a failure to behave according to the light of Reason.

From Hierarchy to Harmony

Parallel to this shifting landscape stretching from Chain discourse to the Enlightenment, another, subsidiary, way of speaking about animals emerged from the sixteenth century. It drew more directly upon biblical texts, and was founded in covenant, rather than place on the Chain of Being or in Nature; and it emphasised harmony rather than coherence.

Thus, when Pope (1912:1734, I.16) sets out to "vindicate the ways of God to man", he frames his task within the Christian tradition of theodicy, yet starts with the Great Chain of Being, and the loss of coherence when humans misplace themselves. His concern is to show that faith in the existence of God is possible in a world welcoming the dualistic Enlightenment consensus of Nature and Man, natural law and freedom. Despite resisting dominant discourse at key points in his essay by prioritising biblical pericopes over Thomist sources, Pope develops his essay within it.

The poet, John Milton (1667, 1.26), writing some 70 years earlier, had also sought to "justifie the wayes of God to men", but he starts with the creation covenant, and a loss of harmony when humans sin. In the theocentric world of Milton's nonconformist discourse, God created humans in a paradise now lost. Milton's task is not, as with Pope, to show how the existence of God is possible in a dualistic Enlightenment world of Nature and Man; but to show how God can be both just and love. His nonconformist ground-motif of Creation, Fall, and Redemption is a significant shift from the medieval consensus, while resisting dualism.

Nonconformist discourse, subjugated in the modern world, was always subordinate but had an influence far beyond its currency. The rest of this chapter recapitulates the themes identified above from within a nonconformist way of speaking, drawing attention to creation singing together..

Nonconformist Discourse

Inheritance

Nonconformists were not isolated from the discursive changes described above. They inherited both the late medieval Chain discourse built upon Thomism, and the innovations of the Renaissance; later nonconformists such as John Wesley and the calvinist Augustus Toplady also interacted with the Enlightenment.

Chain of Being

They found Thomism to be an unreliable guide on matters concerning the Chain of Being, ladder of perfection, and ethic of coherence, as much as on questions of authority and faith. They simply did not find such things in the biblical texts. There is an order to creation, but it is not a hierarchy in the perfection of beings; rather it is a diversity of meaning, each creature disclosing and praising according to its kind. Moreover, the Chain hierarchy was too static for the world they saw in the biblical texts which exhibited an interactive dynamic from the beginning. Creation holds together, not by the coherence of a hierarchy, but by fiat, by the Word, in a harmony of manifestation and praise (Col. 1.15–20). This does not imply that there was a clean break from Thomism. Those most schooled in Greek scholarship found the transition hardest; often it was inconsistent and partial. Nevertheless, a new, nonconformist discourse of inter-relationships between animals emerged, rather than speaking of logic and hierarchical sequence.

This does not imply that nonconformist discourse abandoned the language of the Chain of Being altogether; indeed, some appropriated it and transformed it within their own subordinate discourse of harmony. For example, John Arrowsmith appropriated the language of hierarchy to talk about the way we think about the world in his *A Chain of Principles* of 1659. Yet he presents an "orderly concatenation" among "theological aphorismes and exercitations", rather than a hierarchical ordering of relations among creatures themselves. Watson (1692, 67, s.3) appropriates the concept of an orderly "frame", disordered and rendered incoherent by sin. But his dominant trope is one of beauty and harmony: "Did God make this World full of Beauty and Glory, every thing very good, then what an evil thing is sin, that hath put out of frame the whole Creation. Sin hath much eclipsed the beauty, sowrd [sic] the sweetness, and marred the harmony of the World". John Bulwer (1653, introd., B6) uses "chaine" to gloss "harmony": it is not lawful "to destroy any one species of God's creatures, though it were but the species of Toades and Spiders (because this were a taking away one linke of Gods chaine, one note of his harmony)".

More commonly, however, nonconformist discourse simply bypasses the language of a Chain of Being altogether, drawing upon more biblical sources.

Renaissance Man

Just as nonconformist discourse inherited the Chain tradition, so it also inherited that of the Renaissance. On the one hand, nonconformists drew upon Renaissance scholarship in their return to the biblical texts in the original languages (McGrath 2011, 30). On the other, they had little sympathy with the Renaissance orientation towards the humanities, a trend which would later become known as Humanism. Nor did they value the resurgence of classical literature which portrayed the gods as dependent upon humans for services which they would otherwise lack: upon men to provide food at the banquets of the gods; and upon women for sexual[2] exploitation. As we have seen, nonconformist discourse was consistently theocentric rather than anthropocentric; and it smoothly appropriated the fruits of the humanities such as language study, grammar, rhetoric, and history as gifts of God. The classical stories of the

[2] E.g. Leda, Io, Callisto, Alcemene, Antiope, Europa, and Danaë. These became popular themes for Renaissance painters.

gods creating humans to service divine needs, presented a more specific challenge and demanded refutation.

God's sovereignty and self-sufficiency has a long heritage. God did not, wrote Augustine (1972, XI.24), "create under stress of any compulsion, or because he lacked something for his own needs; his only motive was goodness". God is self-sufficient, but humans are not; we are dependent, we have needs. Nonconformists upended the classical stories of humans servicing the needs of the gods; rather, God provides for human needs. "But why did God make the World?" asked Watson (1692, 66–7), and replies: "Not for himself; he did not need it, being infinite". Rather, the "world was not made for the Eternall God, but for Man", says Thomas Adams (1862:1629, 120). George Walker (1641, 50) elaborates: the world was created "onely for us, and for our benefit, who live under time, and not for the eternal God, to adde any good, or any blessedness to him, who was all-sufficient and most blessed in himselfe from all eternity". God could not gain "any good, or any blessedness" but humans could. When "God created the sunne", preached Calvin (1583, 877), "it was not to lighten himself, but to give light to us"; God is always, already light.

Modern commentators have tended to take phrases such as "onely for us", "for our use", or "for our sake" in isolation from the contrasting "and not for the eternal God", as though such phrases implied that creation is exclusively for humans. But, as we have seen, creation is to manifest and praise the creator; it is not for human consumption. Indeed, it was an act of human (over)-consumption that inaugurated the Fall. God's provision for humans is not at the expense of his provision for all His creatures, as though generosity were a zero-sum commodity. Other animals, like humans, are dependent and also have needs. Adams (1862:1629, vol. 3, 120) speaks of creation as a house made ready before "the Tenant that should have the house": "He creates beasts, but first hee made herbes and grasse to feed them. Hee created man, but not till all things were made fit to entertaine and sustaine him". God made provision for all sensate creatures, animals and humans alike. According to George Abbot (1613, 608), God also had an eye for the world as a whole; He made cattle "to beautifie the great frame of the world, that the earth should not be solitary and naked in any place". The interconnections of sustainability and pleasure suggested here are closer to what we now call an ecological system than to an anthropocentric hierarchy. By reading *for our sake*, or *for our use* within the modern dominant discourse of property relations, such phrases have been transformed into a kind of anthropocentric instrumentalism, as though nonconformists had read Kant. But most had never heard of Kant, not least because they were dead before Kant was born.

Covenant

Nonconformists prioritised biblical text over Church tradition, medieval philosophy, and Renaissance humanism. They did not find in Scripture the hierarchy of Chain discourse, nor the Renaissance dependence of gods on human services. Rather, they found accounts of a people group, and of individuals living lives not unlike their own. Relationships between land, animals, and humans were described in the familiar terms of everyday life: survival in the face of scarcity; and farming practices including animal domestication. They also found there an organising motif for their discourse: that of the covenants which God makes with creation, covenants providing a framework for the mutual dependence of people, animals, and land (Seidenberg 2019, 184). They recognised in these covenants their own pre-modern, agricultural community, circumstances very different from that of modern readers.

I described the Edenic and Noahic covenants in the last chapter; the former dealing with creational norms, and the latter with the consequences of sin. In the following chapters, I will discuss their recapitulation and extension in the Mosaic covenant, especially their provision for the mutual dependence of humans and animals. Three features of these covenants are important for the rest of this chapter.

Creator, Not Creature

Firstly, there is no mutual equivalence between God and creation. This does not mean that there is no relationship. As we have seen, creatures manifest God, each according to its kind; and living creatures call upon the God who hears them, and provides for them. "The End, why all is made, is for the glory of the Maker", says Adams (1862:1629, vol. 3, 120). Not "as if God would purchase a glory he had not before" but to manifest it to us. The "supreme end was his owne glorie; the inferiour and dependant, our benefit and comfort", for "God must be glorified in all creatures, because hee is the maker of all creatures". The ontological, says Pickstock (2007, 58), is "self-exceeded as doxological, as praise of the divine".

Diversity, Not Hierarchy

Secondly, as living creatures, humans and other animals are not placed in a hierarchical order, but share a common calling. In the last chapter we saw many examples of the diversity of creation praising God, insects, fish, mammals, and humans, each according to their kind. Humans and other animals have an ontological continuity as creatures, and a shared solidarity in bodily flesh, but a diversity of voice as singers (Muddiman 1998, 32).[3] This diversity did not reflect a hierarchy of perfection, nor that one creature was created for the sake of another. Genesis, Chapter 1 had clearly pronounced all non-human animals "good" without gradation, fatal to Chain discourse. It portrays a common community of manifestation and praise to God. Moreover, nonconformist diversity was not simply of kinds, but also of individuals within kinds, something unknown to Chain discourse. God not only preserves sparrows, but cares about each individual sparrow.

Co-creaturely Inter-dependence

Thirdly, despite their diversity, there is a mutual inter-dependence between all created beings, especially between the land and all animals; and among all animals, including humans. This inter-dependence ties human identity to that of every other living creature, and our narrative is intertwined with theirs. Humans and land animals were created on the same day, received the same blessings, share the Sabbaths, often bear the same consequences of human sin, and both are found in the covenant community at the restoration of all things. Adam did not identify a specific desire towards Eve as a companion until he had encountered all the other animals, enquired of them, and named them according to their attributes (Moritz 2009, 135–6). Humanness depends upon animals, not as though we are hierarchical intermediaries with angels in a chain of being, nor as Other to irrational creatures, nor by our exercise of power over them, but as co-creatures (Stone 2018, 29–30, 173–4, 180–1).

[3] In both Hebrew and Greek, flesh is the same word for humans and all other animals.

The Harmony of Creation

Nonconformist discourse provided points of resistance to the hierarchical distinctions of their day. It also resisted dominant readings of the human-animal relations as either property, here to serve human needs; or as inherently exploitative, and best kept disengaged. But if it is certain that there are numerous diverse singers, how do they join together in song? How does the unity of the one song emerge from the diversity of the many singers? The harmony of creation will be the focus of the remainder of this chapter.

Harmony and Praise

For Chain discourse, the coherence of the whole is founded in a hierarchy of entities: as Thomas (1924, II.XLIV) says: The best thing in creation is the "perfection of the universe which consists in the orderly variety of things" each in its proper place (Lovejoy 2001, 75). For Enlightenment discourse, the universe finds its unity in instrumental Reason. Dominant discourse has built upon both the hierarchy which authorises property ownership, and the instrumentality which allows animals to be treated as means to an end. Nonconformist discourse rejected both Chain and Enlightenment accounts, proposing a more dynamic interaction of love binding all together in a harmonious whole. There is "variety in unity, and unity in variety'" says Charnock (1864, vol. 2, 22).

The creation sings praise to God, and all song calls for a response, for us to join in; melody contains within itself the basis for harmony. Thus nonconformist discourse reads the creation narratives as presenting a growing harmony, not a hierarchical scale of perfection. The plants and animals are indiscriminately "good". The harmony of diverse voices, of "every thing that he had made" is "very good", a creational community, perfect in praise and manifestation (Gen.1.31). For Thomas Adams (1862:1629, vol. 3, 118, 134), goodness refers to "comeliness and beauty", their harmony in obedience to God, and "commodious and delightfull benefit of them all to man". "That vita vitalis [vital life] requires a confluence of many good things". Its unity is found, not in ownership or instrumentality, but in the harmony of a complete song. It is a tale of dependence, not self-sufficiency; of inter-dependence, not place in a Chain; of orchestra or choir, not in the hierarchy of an army or industrial corporation (Northcott 2020). Conspicuously, the qualifier “good”, let alone “very good”, is withheld from humans in Genesis chapter one.

The nonconformist understanding of harmony respects the individual interests of creatures, framing each within the common purpose of creation: the manifestation and praise of God’s glory. Unity is not vested in *being*, much less in *human* being, but in the *meaning* of creations’ shared task. Fulfilling this task is simultaneously a satisfaction of individual meaning, and contribution to the good of the whole.

The nonconformists did not invent this discourse; it had a long pedigree, not least in the biblical texts themselves. Psalms 148–150 describe a choral work of spiritual, heavenly, and earthly creatures, united in praise. It is also found in other subordinate discourse which prioritised biblical texts. The church father, Athanasius (1971, 117, 115), appropriated and transformed Greek thought, speaking of the Logos (the “Word” of John 1.1) “holding the universe like a lyre” with its differently pitched strings “producing in beauty and harmony a single world and a single order within it”, and of givinglife “to each individually and to all together creating an exquisite single euphonious harmony”. Augustine (1972, 12.5) speaks of “all nature’s substances [having]... in their fashion, a peace and harmony among themselves”. But, if they did not invent it, the nonconformist enriched and elaborated this way of speaking as they made it their own.

The Communion of Creatures

Creation is first a unity. Creation’s voice, says Joseph Caryl (1670, vol. 4, 205, 207), is a “common cry and consent of all creatures”, “there is an intercrying from the lowest to the highest, till the cry come up to the most high God”. The “whole creation”, says Matthew Henry (2008, 1710, Rom. 8.17–25), “the compages [a unity of many parts] of inanimate and sensible creatures, which, because of their harmony and mutual dependence, and because they all constitute and make up one world, are spoken of in the singular number as the creature”. Godfrey Goodman (1622, 21) argues that animals and humans are able to provide tongues for each other in praising God: “Our praise becomes theirs; and their praise becomes ours”. “There is not onely a communion of *saints*, but a *communion of Creatures*, which joyne together in one naturall service of God".[4] This is so prominent a theme that the historian Keith Thomas (1983, 278) has argued that the "modern idea of the balance of nature ... had a theological basis before it gained a scientific one".

This creaturely communion is a kinship of love, peace, concord, and mutual preservation. Spurgeon (1885, Ps. 148.7–10) approved John Everard's image of a family: the psalmist exhorting creatures to praise God sets forth "the sweet harmony that is among all God's creatures; to show how that all the creatures being God's family, do with one consent speak and preach aloud God's praise". Creatures are one family by virtue of creation, with the creational norms of mutuality and love. To describe the interweaving of each creature in this family of creaturely praise, nonconformist discourse employs tropes drawn from music and tapestry.

Music and Tapestry

The last chapter included many examples of creatures as solo voices praising God. This readily extends to the harmony among voices singing together; creation exists as the concordant, intimate relationships between creatures. Thus Everard's "sweet harmony that is among all God's creatures" was a common trope. Martin Luther (1960:1538, vol. 2, 74) refers to a prelapsarian "harmony not only among men but also between the dumb animals and men". Benjamin Parker (1735, 31) eulogised on the prelapsarian friendliness of animals now at enmity with one another; "a perfect harmony and beauty apper'd thro' the whole creation".[5]

An anonymous Puritan wrote of joyful marital love as of the relationship between "two musical instruments rightly fitted [that] do make a most pleasant and sweet harmony in a well tuned consort" (Frye 1955, 155–56). William Cowper (1849:1798, vol. 1, 34, 77–78) uses the same trope of the departed in Christ: "Love makes the music of the blest above, Heav'n's harmony is universal love". The association of music with both joy, harmony, and love, fitted it to describe the communion of praise. For Arrowsmith (1645, 23), the "whole world is a musicall instrument;

[4] Goodman converted to Catholicism, probably in the 1640s. However, his theology of creation of 1622 was commended by C. H. Spurgeon (1885, Ps. 148.7–10) as compatible with the nonconformist discourse as regards creation.

[5] Parker appears to have been a self educated layman who applied nonconformist perspectives for a general public, at least so far as the doctrine of creation is concerned.

every string is so tuned, as to be fitted for resounding the great Creator's prayse".

There is one choir but many voices, each contributing its part to the song of praise; one consort but many instruments. The harmony, the interweaving of parts, forms the whole. This language derived from biblical pericopes, especially the psalms, and has been used by some modern theologians. Creation, says Daniella Rizzo (2024, 70), transforms "into an orchestra, where every element contributes its unique note to the symphony of God's praises. This symphony, in turn, extends an open invitation to human participants".

As well as the musical tropes which echo biblical pericopes, nonconformist discourse also drew on the art of tapestry or marquetry; like musical harmony, these emphasised interwoven-ness or inlaid pattern, over the single thread or linear grain. The world, said Thomas Watson (1692, 67), "is like a curious Piece of Arras or Tapestry, in which we may see the Skill and Wisdom of him that made it" Stephen Charnock (1864, vol. 2, 26) preached about a world "interwoven and inlaid together", "one entire beauty in the whole fabric". Like music, tapestry and marquetry portray a harmony of relationships, not connected in a one-dimensional chain, but interconnected on a picture plane. Rather than *being* the song of creation, it portrays that song, as a musical score portrays a symphony to the musician.

A Common Centre

The harmonious song of manifestation and praise is simultaneously a satisfaction of individual and relational meaning. It also expresses the good of the whole. The fabric of creation meets "in a common centre, the good and preservation of the universe", says Charnock; "All creatures are as members in the great body of the world, proportioned to one another, and contributing to the beauty of the whole…".. For Adams (1862: 1629, vol. 3, 118), every individual part is "enbued with speciall vertue" suited to it's purpose within the whole creation". All creatures, says Matthew Henry (2008:1710, Ps. 104.19–30), were made "to answer the end they were designed to serve, the good of the universe… to the glory of [God]".

The Human Place

In this chapter we have seen that creation as praise draws our attention to diversity in unity, to individuals in harmony, and to the good of creation manifesting the glory of God. It does so in a way that abundantly exceeds its material presence, inspiring an awe which might even threaten to undo us. This resists at numerous points the dominant understanding of animals as material property, or even the subsidiary discourse of living beings.It also raises the question addressed in the next chapter: what, if anything, does this sense of surplus over what is necessary for survival demand of humans?

References

Abbot, George. 1613. *An Exposition upon the Prophet Jonah*. London: Richard Field.

Adams, Thomas 1862:1629. Meditation Upon Some Part of the Creed. In *The Works of Thomas Adams, vol.3,* 85–264. Edinburgh: James Nichol.

Alexander, C.F. 1848. All Things Bright and Beautiful. In *Hymns for Little Children,* 27. Philadelphia: Herman Hooker.

Arrowsmith, John. 1645. *Englands Ebenezer*. London: Samuel Man.

Arrowsmith, John. 1659. *Armilla catechetica. A Chain of Principles; or, An Orderly Concatenation of Theological Aphorismes and Exercitations...* Cambridge: John Field.

Athanasius. 1971. *Contra Gentes and De Incarnatione,* ed. R.W. Thompson. Oxford: Clarendon.

Augustine. 1972. *City of God,* trans. Henry Bettenson, Penguin.

Bulwer, John. 1653. *Anthropometamorphosis.* London: William Hunt.

Calvin, John. 1583. *Sermons on Deuteronomy,* trans. Arthur Golding. London: John Harrison.

Caryl, Joseph. 1670. *An Exposition with Practical Observations upon the Book of Job*. London: Thomas Sawbridge.

Charnock, Stephen. 1864. *Discourse on the Existence and Attributes of God. In Works,* vol 2. Edinburgh: James Nicol.

Cowper, William. 1849:1798. The Progress of Error. In *Poems,* ed. John Johnson. London: Phillips, Sampson & Co.

Cowper, William. 1967:1785. The Task.... In *The Poetical Works,* 129–241. Oxford: OUP.

Damrosch, Leopold. 2005. *Jean-Jacques Rousseau.* Boston: Houghton Mifflin Company.

Frye, Roland. 1955. The Teachings of Classical Puritanism on Conjugal Love. *Studies in the Renaissance*, 2.
Goodman, Godfrey. 1622. *Creatures Praysing God or the Religion of Dumbe Creatures.* London: Felik Kingston.
Hegel, G. W. F. 2019. *The Phenomenology of Spirit*. University of Notre Dame Press.
Henry, Matthew. 2008:1710. *Commentary on the Whole Bible.* Peabody, MA: Hendickson.
Kant, Immanuel. 2009. *An Answer to the Question What Is Enlightenment?* London: Penguin.
Kant, Immanuel. 2000. *Critique of the Power of Judgment*. Cambridge University Press.
Kant, Immanuel. 1930. *Lectures on Ethics,* trans. Louis Infield. London: Methuen.
Lovejoy, Arthur O. 2001. *The Great Chain of Being*. Cambridge, Mass: Harvard University Press.
Luther, Martin. 1960:1538. *Works,* eds. Jaroslav Pelikan. Saint Louis: Concordia.
McGrath, Alister. 2011. *Christian Theology*. Oxford: Wiley-Blackwell.
Milton, John. 1667. *Paradise Lost.* London: Peter Parker.
Moritz, Joshua. 2009. Animals and the Image of God. *Dialog: A Journal of Theology* 48 (2): 134–146.
Muddiman, John. 1998. A New Testament Doctrine of Creation. In *Animals on the Agenda*, ed. Andrew Linzey and Dorothy Yamamoto, 25–32. London: SCM.
Northcott, Michael. 2020. Covid 19, Human Ecology, and the Ontological Turn. In *Pandemic, Ecology and Theology,* eds. Alex Hamilton. London: Routledge.
Parker, Benjamin. 1735. *A Second Volume of Philosophical Meditations.* Mary-Bridge Derby: Parker.
Pickstock, Catherine. 2007. God and Meaning in Music *Sacred Music*, 134 (4): 40–62.
Pico della Mirandola, Giovanni. 1956. *Oration on the Dignity of Man*. Gateway editions.
Preece, R., and Fraser, D. 2000. The status of animals in Biblical and Christian thought. *Society and Animals* 8 (3), 245–263.
Pope, Alexander. 1713. Against Barbarity to Animals. The Guardian, 61, 21 May.
Pope, Alexander. 1720. Windsor Forest. In *Miscellaneous Poems and Translations,* 7–26. London: Bernard Lintot.
Pope, Alexander. 1912:1734. *Essay on Man*. Cambridge: CUP.
Rizzo, Daniella. 2024. Animal Glossolalia: A Pneumatological Framework for Animal Theology. *Pneuma* 46: 60–79.

Seidenberg, David Mevorach. 2019. Veganism and Covenantalism. In *Jewish Veganism and Vegetarianism*, ed. Jacob Ari Labendz and Shmuly Yanklowitz, 161–194. New York: Suny Press.
Solomon, Harry M. 1993. *The Rape of the Text*. Tuscaloosa: University of Alabama Press.
Spurgeon, C.H. 1885. *The Treasury of David*. London: Marshall.
Stone, Ken. 2018. *Reading the Hebrew Bible*. Stanford University Press.
Thomas Aquinas. 1924. *Summa Contra Gentiles*. London: Burns and Oates.
Thomas Aquinas. 1947. *Summa Theologica*. Notre Dame: Christian Classics.
Thomas, Keith. 1983. *Man and the Natural World*. London: Allen Lane.
Walker, George. 1641. The History of the Creation. In *God Made Visible in his Works*. London: Bartlet
Watson, Thomas. 1692. *A body of practical divinity* London: Thomas Parkurst.

CHAPTER 5

The Art of Conducting

Introduction

The previous two chapters have discussed the songs of individual animals, and singing together. This chapter addresses the place and office of humans in this cosmic song of praise.

What is it to be authentically human? For the dominant discourse, it is to be at the centre of the world, sovereign over animals. For modern subordinate discourses, to be authentically human is to recognise our place as an animal among animals. Nonconformist discourse rejected both alternatives. It considered that to be authentically human is neither to be head of a hierarchy, nor to be absorbed into the unity of one among many, but to be a fellow musician, to conduct creation's song.

Human and other animals have an ontological continuity as dependent creatures, and a diversity of voice as singers. Humans can partner with the diverse voices of non-human others, in a co-creation and co-sustaining of song (Northcott 2020). This opens a space in which a novel ethic towards animals can consolidate. One that shifts the emphasis from a disorderliness in the Chain of Being, or a lack of rationality, to the particular animals themselves, and the human duty to enable all creatures to bring glory to God, each according to its kind.

P. J. Sampson, *Nonconformist Perspectives on Animals and Language*, The Palgrave Macmillan Animal Ethics Series,
https://doi.org/10.1007/978-3-032-12158-5_5

Yet nonconformists were not blind to the cruelty they saw all around them. How, they asked, do humans conduct a suffering choir? That question will be addressed in Chapter 5.

Human-Animal Relationships

All human discourses about animals reserve a place for humans. In modern dominant and subordinate discourses, human being is autonomous and privileged. A shared question inaugurates both discourses: how do humans fit into the world? What is their place? Is it to be at the centre of the world, sovereign over everything else? Or is authentic humanness to be an animal among animals?

I have argued that nonconformist discourse speaks about animals differently. Along with all creation, animals find the meaning of their existence in the manifestation and praise of God.

As feminist and post-colonial theorists have noted, our discourse of being human supports and is supported by the way we speak about the Other. What it is to be human is decisively affected by the way we speak about animals (Sampson 2010; Stone 2018, 24f, 126f). When nonconformist discourse speaks differently about animals, this changes what is meant by "human".

In nonconformity, humans are neither self-sufficient and privileged, nor simply equivalent to all other animals. To be authentically human is neither to be head of a hierarchy, nor to be absorbed into the unity of one among many. Rather it is, as a fellow musician, to sing according to our kind alongside others in "the feast of praise". All animals have their own voice; they praise God according to their kind, the human no less and no more than the others. What, then, is the distinctive human voice?

Singing Together

For good or ill, contemporary discourses about animals begin with humans and with human behaviour towards other animals. Moreover, humans are represented as either disconnected from other animals, or as disconnecting themselves. In the dominant discourse, this is portrayed as human privilege and superiority. In subordinate discourse, despite our kinship as living beings, there seem few options for significant connections with other animals which do not exploit them; free-living animals are better off without human contact. Humans either tyrannise commodified animals, or break off all relations for fear of tyranny.

Nonconformist discourse begins with God, not humans. Humans and other animals alike are creatures, serving the same end: the manifestation of God; and singing His praise. There is not one fundamental disconnection between humans and all other animals, but multiple differences, according to kind (Stone 2018, 74–80). Human culture is no longer isolated at the centre of things, autonomously generating representations of all other animals and our reactions to them, but exists in a common space with, as Derrida (2008, 9) puts it, the "un-substitutable singularity" of individual animal cultures. Discourse about animals is predicated upon an engagement *with* them in praise, not observation *of* them. Where the dominant discourse sees human-animal relations as proprietorial, and subordinate discourse as exploitative and best avoided, nonconformist discourse pictures a horizontal gaze of wonder, joy, and glory; a community of jubilant creatures on an equal ontological footing.

This lack of both human self-sufficiency and privilege is exemplified by the four living creatures of Rev. 4.6–9 who continually offer "glory and honour and thanks" "round about the throne" of God. Representing animal creation in the categories of the day, they declare their dependence upon God in a harmony of voices, a "collective language that transcends conventional verbal communication" (Rizzo 2024, 69). The pericope describes four categories of animal life: free-living, domesticated, human, and avian. One creature is "like a lion", the king of free-living creatures; another, "like a calf", stands for domesticated animals; the third with "a face like a man", represents humans; and finally one "like a flying eagle", is cardinal among birds. The humanlike creature is simply one among the four, not given precedence or even listed first (Bauckham 2008, 62; 2011, 221–22. Wright 2011, 47–8). Indeed, the creatures' arrangement "round about the throne" suggests a symmetry which would exclude all priority. They share creatureliness, singing together, not with cool appraisal, but passionate, joyful acclamation.

Sharing and Teaching

Co-creatureliness in praise implies a mutuality of embodied relationships. This includes a background of shared capacities. All creatures, says Lancelot Andrewes (1650, 315), "love one another; one beast delighting in another of the same kinde, and little children delighting in their image in the glasse, shew this". Baxter (1670, 184–85) regarded all equally as epicures: "A beast tasteth as much of the sensitive sweetness of his food and ease as you do". More specifically, humans and other animals are bound together "by the lawes of the grand Creatour", says George Abbot: "there is such affinitie betweene man and the beastes, which are subjected to his use, that the sorrowes of the better do easily touch the worser. For God hath so coupled all creatures to mankind, with a chayne of strong dependance, that the being of them is much sutable to the flourishing, or fading of the other". So intimate is this co-agency that humans and animals, flourish, fade, fall and are restored together: "by the fall of our first parents, [animals] fell into grievous bondage, even sinking in their excellencie, when man did sinke..." "In the day of judgement... [t]hen shall they returne to that beautie wherein they at first were established..." (Abbot 1613, 450). Humans and animals share agency, vulnerability, mutual interests, and hope (Stone 2018, 113–14, 146). The mutual relationships of living creatures is reflected in the vocabulary of the biblical texts; in both Hebrew and Greek, the word for "soul" or "self" is the same for both humans and other animals; the same applies to the word for their "flesh" (Seidenberg 2019, 164).

This community of jubilant acclamation is exemplified by John Milton (1667, VIII.514–520) when he portrays the birds sharing the celebration of the marriage of Adam and Eve. On each hill: "Joyous the birds; fresh gales [outcries] and gentle airs [melodies]

> Whispered it to the woods, and from their wings
> Flung rose, flung odours from the spicy shrub,
> Disporting, till the amorous bird of night
> Sung spousal, and bid haste the evening star
> On his hill top, to light the bridal lamp".

The modern reader sees here figurative language, but as I discussed in Chapter 1, nonconformist discourse encompasses more substantial relationships between word and thing.

School and Fall

This background of mutuality was disrupted at the Fall. Animals fell into "grievous bondage" but were not themselves fallen in the same sense as humans. Where humanity was compromised in its ability to praise, animals were less so, and could even take the lead. God, says King (1864,1594, 267), has enabled beasts and birds to glorify their Maker "with strength and comeliness of nature more than ourselves". Animals can even be voice coaches to humans; they can teach us how to sing. If, says John Calvin, a man "wishes to be considered a reasonable creature, having mastery over the brute beasts... [let him] be sent to school to the beasts" (Bouwsma 1988, 140). "All creatures" says Caryl (1670, vol. 4, 206, 211), "have a teaching voice, they read us divinity lectures of a Divine Providence". Each "shall teach thee, every one by himself shall teach thee, the least as well as the greatest, the Mouse as well as the Elephant or the Lyon; the Shrimp as well as Leviathan... So many creatures as are in the world, so many Preachers there are in the world, of the power and goodness of God". One fewer animal, one less voice, a lesson lost. Animals are, says Ralph Venning (1669, 75–6), "schoolmasters to man": "if you are a prayerless or distrustful person these creatures witness against your sin, for they teach you to pray and trust". Calvin (1583, 776) bids us imitate a mother bird he had seen who had "such a care of her young, that she forgetteth herself for their sakes"; when birds "see their little ones, to their seeming they be deerer to them than their owne life". When Jesus of Nazareth compared himself to a hen with her chicks, he assumed that the crowds, like Calvin, had seen this, and learned the lesson. He recognised the hen's love, taking this as the standard for his own compassion for the city of Jerusalem.

In 1686, John Bunyan (1980:1686, 241, 214–221) published *A Book for Boys and Girls* which contains, among other things, instances of animals educating us in the first person. We find spiders, ants, and other lowly creatures, rebuking us, should we have trouble accepting their advice. The poem *Upon the Pismire* [ant] tells us that we must "unto the Pismire go to school"; "Obey, her ways, consider, and be wise. The Pis-ants tell thee what thou must do, And set the way to Life before thine eyes". In *The Sinner and the Spider*, the sinner despises the spider as "scum" and "dross", threatening to squash it. The spider replies that he has the unquestioned right to live in even a king's palace. Moreover, the "ugly" spider "keep[s] the rules of [his] Creation" while the sinner is made uglier than the spider by his sins, even though made in the image of God. The spider has integrity, and the sinner finally admits "Thou art my Monitor, I am a fool; They learn may, that to Spiders go to School".

As Milton's portrayal of the birds celebrating with Adam and Eve demonstrates, mutuality preceded the Fall; there seems no good reason to suppose that the teaching role of animals did not also. Certainly, the animals set an example, as the case of "joy in praise" illustrates.

Joy

We enjoy God which is, says the Westminster Shorter Catechism (1647, Q1), a "chief end of man"; and we enjoy God manifested in embodied creatures. Adam, says John Wesley (1872:1771, para. 2.I.2), "saw, with unspeakable pleasure, the order, the beauty, the harmony, of all the creatures; of all animated, all inanimate nature; the serenity of the skies; the sun walking in brightness; the sweetly variegated clothing of the earth; the trees, the fruits, the flowers, and liquid lapse of murmuring streams... Nothing could stain his purity of joy". "Being placed in this most beautiful theatre", says Calvin (1845,1559, 1.14.20), "let us not decline to take a pious delight in the clear and manifest works of God". Here, "pious" does not imply ethereal or "other-worldly". God "created food, ... not only for our necessity, but also for our enjoyment and delight. ... herbs, fruits, and trees, besides their various uses, gracefulness of appearance and sweetness of smell". Calvin employs specific terms for "enjoying" and "taking delight" in creation, distinct from its pragmatic utility for our survival (Wevers 1992).

As we have seen, animals also delight in one-another, enjoy the taste of food, and cherish their leisure. The "birds sing in the air, the beasts skip in the pastures", says Adams (1862: 1629, vol.3, 203, 125). They teach us by example: "How can we be mute when we hear the little birds every morning carol the praises of their Maker?". Anne Bradstreet (1962:1650, 373, v9) contemplates the same theme: "I heard the merry grasshopper then sing. The black-clad cricket bears a second part; They kept one tune and played on the same string, Seeming to glory in their little art. Shall creatures abject thus their voices raise And in their kind resound their Maker's praise, Whilst I, as mute, can warble forth no higher lays?". Jamieson et al. (1871, Mal. 4.2) compares the humans freed from captivity with the "calves of the stall - which, when set free from the stall, disport with joy". There is nothing ethereal about such joy. It is embodied in grasshopper and calf, skipping and singing, in sound and scent, in eating and drinking, which other animals enjoy as much as we do.

Humans share co-creaturely relationships with other animals in which they sometimes take the lead, as school-masters. But there is also a distinctive place for humans according to their kind. This is often referred to as *dominion*, or as being created in the *Image of God*.

Dominion and Image

As we saw in Chapter 1, when *dominion* and *Image of God* are appropriated by the dominant discourse they are transformed into tyranny, instrumental use, and proprietorial ownership. Subordinate discourse, seeing little alternative, accordingly favours disengagement between humans and other animals to avoid this possibility. In these appropriations of dominion, the creation of humans on the "sixth day" (Gen 1.24–31) is read hierarchically as the culmination of creation, despite the fact that the land animals were also created on the sixth day. Admittedly, humans were created after land animals, in the afternoon as it were, but if this implies precedence, it is familial not hierarchical; animals are "our elder brethren: for their creation was previous to ours" as Augustus Toplady (1825,c1778, 462) put it.

Dominion

Within nonconformist discourse, *dominion* and *Image of God* occupy entirely different places from those in the story of the dominant discourse. Human dominion over the other animals does not endow ownership rights, nor does it permit man "to doe with the creatures what he will, and to use them as hee listeth" (Walker 1641, 227). Humans are on the same footing as a fellow musician. As Joseph Hall (1654, 289) wrote: "God made man the lord of his Creatures; he made him not a Tyrant". This lordship consisted, says Luther, not as we use the creatures today, but "principally for the admiration of God and holy joy" (Linzey 1987, 13).

All nonconformists considered that the Fall changed this, and most agreed with Luther that dominion implied a co-operative relationship between humans and animals which is now almost entirely lost. We now have, said Luther (1958, vol.1, 67), "an extremely small and far inferior" dominion than Adam. Indeed, "the name and word 'dominion' [is] as a mere title, but the substance itself has been almost entirely lost", although it will be "restored to us" in Paradise. Presumably he had in mind the pericopes prophesying that the "beasts of the field shall be at peace" with men, and that children will suffer no harm from now dangerous animals (Job 5.23, Is. 11.6–8). Jeremiah Burroughes (1643, vol. 1, 576) agreed that we "have lost by sin a great part of our dominion that God gave us over his creatures, that was the result from that Image of God that man was created in". However, the partial loss of our peaceable dominion did not remove our responsibilities arising from being in the Image of God.

Image

In the Ancient Near East, the *image of God* was a complex cultural emblem, including the fabricated image of the God in the temple; and the living image of the God as the king. Nonconformist discourse drew on the latter, picturing humans as the "kingly representative" of God on earth, although the manner of this representation varied; sometimes it is seen as an identity manifesting attributes of God as human faculties; sometimes as a calling to have dominion over the earth; often as a mixture. Here, it is sufficient to note that the narrative of human creation connects the *Image of God* with the function of *dominion* as we have just seen Burroughes intimate (Berkouwer 1962).

The biblical text states that humans are made in the image of God (Gen. 1.26–28), and nonconformist discourse avoids describing animals in the same terms. Wendy Doniger (1999, 100, emph. Orig.) observes that this is unusual; "most mythologies assume that animals, *rather than humans*, are the image of god - which may be a reason *to eat them*", thus acquiring the god's qualities through consumption. That only man is made in God's image thus emphasises that, in the absence of anthropophagy, flesh is off the menu.

In nonconformist discourse, the human image of God is unique but not exclusive. Since all creation manifests God according to their respective kind, other creatures also image God in some respect, though are not themselves the image of God. As George Walker (1641, 265–7) points out, "other creatures have more or less resemblances and impressions of God, some even excelling human imaging in some respects". This is especially so since the Fall when the Image of God in humans was lost or "at least grossly corrupted and distorted". The modern appropriation of *Image of God* transforms postlapsarian distortion into perfect representation, and annuls the capacity of other animals to also image God.

For Walker, as for most nonconformists, it is the Image of God which made humans fit to have dominion, and to rule according to God's will. Originally, the Image included perfectly understanding "the nature of all the creatures of God, and what was good both for himself and for them, … and what was just for him to do: and how to beare himself uprightly towards God and all his creatures". This concern for the good of the creatures, and for justice and uprightness towards them, enabled man "to rule all Creatures with discretion", to be "able to will just things justly", and to be "delighted in the good creatures of God". Like the dominion which it permitted, the Image was to a greater or lesser extent lost at the Fall, or at least grossly tarnished and distorted. Since it is the Image of God which was to have enabled man to exercise dominion, we must understand how God reigns over creation in order to appreciate how man, as His image, was to have done so, and how we might do so again.

God's rule over creation manifests His love, not instrumental use as with the gods of the classical world. The reason, says Calvin (1855:1551, 4.9), that the world was created, and that we have dominion over it, is the same: "the gratuitous love of God", for "God is love". God the Creator, says Caryl (1670, 207) "betrothed himself in mercies and loving kindness" to all the creatures of earth. This, not in some general sense, as His providential care extends to individual beasts: "Yea, even to the 'calving of hinds', to the 'feeding of lions and young ravens'", says Adams (1862,1629, 165). So humans should image God's care in their dominion; not out of self interest, but for "love to the creature, or him that made it", say Dod and Cleaver (1612, 140–41). "As all creatures doe taste of, and live by the aboundant liberality and bountifulnesse of Gods hand, so he would have them to feele by sense, though they cannot discerne it be reason, that there is also care for them and compassion in his children". Charnock (1864a, 87, 80) compares God's sustaining care with that of a nursing mother (Is. 49.15). As we have seen, Calvin bade his congregation learn tenderness from a mother bird. John Ray (1701, 219) noted that scripture often mentions the 'provision that is made for [animals].. sustenance"; the mountains "serve for the Harbour, Entertainment, and Maintenance of various Animals, Birds, Beasts, and Insects, that breed, feed and frequent there". Robert Boyle (Oster 1984), in a manuscript of 1647–8, found many pericopes showing God's compassion for animals; God calls animals his own creatures "equally with us", and he cares for them; the Sabbath provision for animals implies God's "Tenderness of Brutes". Birds, says Arrowsmith (1659, 416–17), "have no caterers to bring in provision from them; no barns to fetch provision out of; yet want it not, but go cheerfully on, chirping continually, because God feedeth them: and that sometimes in a way little less then miraculous".[1] Human rule, like God's, is not to exploit or tyrannise, but to be the service of the shepherd king, whose subjects merit covenant care (Seidenberg 2019, 164). God, preached Calvin (1583, 877), will have us "nourish them [animals] and to have care of them".

Justice and Uprightness

If human dominion is to image God's care of creation, we should reflect His justice. We should, as Walker said, act with justice and uprightness. Calvin (1583, 877, 775, 770) insists that "as we deal rightfully with men, so we should use the like duty even towards the brute beasts, which have no reason, nor understanding, nor cannot complain of the injuries which are done unto them". To ignore the demand of the dumb is a mark of the wicked. "A young bird hath no tong to crave reason at our hands, yet we are to behave uprightly towards them". Animals, as part of the covenant community, have what we would now call "rights", and humans have duties tojustly provide for them. Boyle's reference to "the Sabbath provision" is a case in point. Sarah Burdett (1839, 17) complained of cruel men who do not allow animals "their just right of rest". But this is not only a right for animals to claim rest from labour. It is also a covenant right to claim the promises of restoration. Thus, according to James Ussher (1653, 248), Sabbath provision represents "after a sort of everlasting Sabbath, wherein all creatures shall be delivered from the bondage of corruption". Or, as Caryl (1670, 208, emph. Orig.) puts it: "there is somewhat in the creature, not only like *expectation*, but like *the earnest expectation*" of deliverance. Animals, it seems, may rightly expect restoration; but certainly they have a right to just treatment. Referring to Prov. 12.10, James Tyrrell (1694, 104) observes: "I doubt not but the very Brutes have a Right to be Governed with mercy andJustice, and that God who is their Creator as well as ours will punish cruel men if they Tyrannize over them".

Rawlinson (1612, 9–10) argues thatjustice towards animals is so tied to human identity as made in the image of God, that to be truly a man is to

[1] By "miraculous", he refers to a then current account of orphaned ravens fed from worms breeding in their dung.

be merciful to animals. Discussing the pericope "a just man is merciful to the life of his beast" (Prov. 12.10), he argues that "every *man* (truly and properly so called) is *just*, yea so just, that if hee bee not just, he is no man: so may I also truly averre, that every *just man* is mercifull, yea so *mercifull*, that if hee bee not mercifull hee cannot be just [and so be no man]".[2] Charles Bridges (1865, 121f) makes the same connection between human nature and the Image of God: "A righteous man regardeth the life of his beast, duly attends to its comfort, and never presses it beyond its strength. The brutal habits therefore, the coarse words, inhuman blows, and hard tyranny on the public roads is disgraceful to our nature. ...But why is this humanity marked as the feature of a righteous man? Because it is the image of our heavenly Father, who spreads his cherishing wings over his whole creation ...Must not then his children reflect his whole image of love?"

As Christ is the lord of all things, "the proper right and interest of them all is in him", says Walker (1641, 161). For Charnock (1864b, 351), the rights of animals are set out in the law; "How should we be careful not to use those unmercifully which God takes such care of in his law...". As well as Sabbath rest, animals are also entitled to food (Deut. 25.4). The "withholding of due and necessary food from the Beast, or overburdening it beyond measure, is a dishonour offered to its Creator", says John Maynard (1688, 19). Men, says Calvin (1958, 329), "are required to practicejustice even in dealing with animals" which "are entitled to their food". As discussed in Chapter 1, this understanding of *rights* or *entitlement* is broader than its modern equivalent (Linzey 1987, 95).

The Art of Conducting

Human dominion, then, entails care and upright dealing towards animals amounting to service. This images God's re-orientation of "rule" which, according to Jesus of Nazareth, is not to "exercise lordship... and authority" over others, but to be "the servant of all" (Mk 10.42f; also Matt.20.27, Phil. 2.7). As a fellow musician, how are humans to fulfil this calling in the creatures' choir of praise? I suggest by taking on the obligations of the conductor.

[2] Rawlinson's affiliations are debated, but here he explicitly states (1612, 1) his intention to follow "Mr Calvin".

The whole world was created for this one purpose: to manifest and praise God its creator, and man is not given dominion until the *end* of the creation narrative. First, the great orchestra or choir is created, then a conductor is appointed. Like mercy, conducting is "twice blessed"—delighting both humans and other animals in a creation now declared, not just "good", but "very good". The orchestra is the Lord's, neither owned nor to be consumed by men. Indeed, immediately after appointing the conductor, God gives all the musicians a vegetable diet (Gen. 1.26–30) which ensures a fear-less song. Conducting is the opposite to consuming. Conducting is to draw out qualities, to make them distinct, to let each voice sound out its praise while remaining part of the whole body of musicians. To consume is to devour the bodies of the musicians, to chew up and homogenise. This already raises the possibility that the conductor will abuse the power of his calling; will tyrannise rather than serve, consume rather than create. But the conductor's *calling* is to serve not consume (Ez. 34.10–12): to get the best from the individuals; to ensure passion, and a balance between voices or instruments; sometimes to call the soloist to sing alone. Human beings uniquely have the capacity to do this, to know God's will and direct other creatures towards their proper end (Linzey 1987, 98). Of course, there are many way to conduct, from the precision of Pierre Boulez to the flamboyance of Leonard Bernstein. But whatever personal style a person may choose, the conductor must know the voices of the other musicians.

On Knowing the Voices

The nonconformist discourse of animals goes beyond their physical care to the "good of the creatures", to a recognition of the individual animal's nature, and to what we might now call their cultural repertoire. Fish, birds, mammals, each have their own domain (Gen. 1.20–25). The animal is not just *the general singular*; each animal community has a culture just as humans do, and those cultures are diverse, with distinctive patterns of social behaviour, language, knowledge, and customs which are transmitted from one generation to the next (Derrida 2008, 408–9. Laland and Galef 2009. Whiten 2021). Within each culture are individual animals. Nonconformist discourse speaks of them delighting in one another's company: they "cherish and tenderly care for their young"; they experience pleasure, know justice, and even have musical and spiritual lives (Tillotson 1700, 35–6. Stone 2018, 142 f. Sampson

2018, 74f)). Such attempts to contemplate the animal's voice beyond anthropocentrism were rare in their day, and are not commonplace in ours.

Keil and Delitzsch (1973, 255) translate Proverbs 12.10 as: "the righteous knows how his cattle feel". The object phrase also occurs in Exodus 23.9 where it refers to knowing how it feels to be oppressed as a sojourner or stranger: "to be sympathetically aware of" another's condition. It elsewhere refers to God's compassion in both the Hebrew, and in the Greek of St.Paul (Rom. 9.15). The righteous man can empathise with what animals are feeling; he knows the animal well, and feels for her as we might feel for a friend. The good shepherd knows the sheep individually, by name (John 10.3). Balaam, who had known his ass for years, lacked just this empathy (Num. 22.23ff), a lack which Matthew Henry concludes leads to animals being "abused or tyrannised over. Balaam was checked [rebuked] for beating his ass" (Clifford 1999, 131. Bauckham 2011, 138–40. Henry 2008:1710, Prov. 12.10).

This is not simply prudent property maintenance, but a compassion which, when it sees cruelty, has our "bowels... stirred within us, ...[and] our hearts be pricked... [like] God, who is very pitiful, and of tender mercies!" (Tillotson 1700, 129). It is the goodness which "sweeps a wide circle touching the throne of God and the stall of the cattle" (MacLaren 1908, 160). It is the duty of the conductor to serve the orchestra, to care for the musicians so they can play at their best. We should, as Walker (1641, 160–1) put it, "respect and esteem all creatures, as the workmanship of God, made and framed by the same hand which formed us; and to take heed, and beware that we do not abuse any of God's creatures to feed our own vain appetite, to satisfy our sinful desires and pleasures, or to serve our corrupt fleshly lusts". Such sensibilities take us a fair way to understanding the calling of humans in creation's harmony. The conductor must know the heart of the musicians.

When Adam named the animals, it was not, according to Luther (1958, vol. 1, 119–20), casual or arbitrary, but after carefully attending to them and arriving "at such a knowledge of their nature that he can give one a suitable name that harmonises with its nature". Likewise, Calvin (1965,1554, 132) emphasises that appropriate names, "agreeing with the nature of each" were given only after "having closely inspected them". This capacity was specific to humans: "All arts were engraven upon the creatures; yet none but man could see them" says John Yates (1622, 159). Of course, this posits a connection between name and nature which is foreign to modern linguistics: "I nam'd them as they pass'd, and understood Their nature, with such knowledge God endu'd My sudden apprehension", says Milton (1667, VIII.352). But it also points to the recognition of natures not our own, knowable only after close inspection or direct revelation. "So then", says Joseph Caryl (1670, 210), "we ask the creatures [for knowledge], when we diligently consider them, when we search out the perfections and vertues which God hath put into, or stamped upon them; to set our minds thus upon the creature, is to discourse with the creature". The righteous knowing how cattle feel, says Alexander MacLaren (1908, 160) "seems to suggest sympathetic interest, leading to an understanding of the dumb creature, which must precede all wise care for its well-being. It is a part of religion to try to enter into the mysterious feelings of our humble dependants in farmyard and stable". The righteous man, says Adam Clarke (1832, Prov. 12.10), considers "what is best for the comfort, ease, health, and life of the beast".

Both MacLaren's "sympathetic interest", and the attention required for Adam to name other animals, bear a family resemblance to what has been called "listening otherwise", an "evenly hovering attention" that suspends comparisons with human experience in order to receive the singularities of the other (Lipari 2009, 44. Laplanche and Pontalis 1980, 43–4). Attention hovers over other animals, as the dove of the Holy Spirit hovered over the face of the deep (Gen. 1.2). As animals reveal themselves to us, they place an ethical demand upon us arising from the other's way of being present. In attending their voice, a demand arises from a kind of listening that *speaks*, that "is awakened and attuned to the sounds of difference rather than to the sounds of sameness". In attending to other animals, we respond to that demand. For Emmanuel Levinas (1969, 207, 187f, 192–3), this attending, prior to understanding, is transcendence; it is revealed to us. If creation is a mode of revelation, there is no lack of speech, only of listening and watching. Anyone who has had any dealings with animals knows that they are neither deaf nor silent. The problem is not that animals lack a voice, but that we possess the human language of advocacy. Out of love, we should listen, starting, as Calvin (1583, 770) observed, by going to the animal "of our own good will, though we be not moved or requested there-unto", to "listen like one being instructed" (Is. 50.4). Only when we have listened and know the voice of the other animals, can we begin to conduct them in the chorus of praise.

On Knowing the Ensemble

The capacity of humans to name other creatures, to learn their nature, is indicative of a wider capacity: to know the reason why. In the musical domain, to know how the different parts fit together, how the overall composition is moving to its finale. This also is the role of the conductor or choirmaster (Wright 2011, 47–8, 50).

To conduct an orchestra is not to be superior to the orchestral players, but to coordinate in a common cause. For Martin Luther (1960:1538, vol. 2, 74), the harmony which God pronounced "very good" in the beginning, was a "harmony not only among men but also between the dumb animals and men". "The whole world", says Arrowsmith (1645, 23), "is a musicall instrument; every string is so tuned, as to be fitted for resounding the great Creator's prayse... man ...is the onely musician that knows how to play upon this instrument". To shift the trope from instrumental to orchestral performance, humans are the only creatures tasked to conduct in love and care the great orchestra of praise and manifestation which is the purpose and fulfilment of creation. When we listen to the voice of our fellow creatures in their desire to worship God, and in solidarity take up our task as only humans can, then we fulfil what it is to be human, to image God. Only when the conductor inspires, transmits, and shares her joy can the orchestra fully celebrate the music. From creation's point of view, this is why it sustains us. This is the human calling; to, as John Owen (1851,1657, 52–3) puts it, "with the Psalmist, call in all their fellow creatures to the work" of exalting God. That is not to say that humans have always fulfilled their task. According to the nonconformists, humans did quite the opposite.

Fracture

The nonconformists were acutely aware that humans do not fulfil their calling; that we use our power to make other animals suffer, even to the point of consuming them. It was sin, said Thomas Adams (1862:1629, 137), "that made us butchers, and taught the master to eat the servant". All around us, says John Wesley (1872:1771, para.2), we see that "a complication of evils oppresses, yea, overwhelms" living creatures. Everywhere they turned, they heard disharmony in the animal creation: predation, starvation, and disease; "vanity under which they are enslaved", as Charnock (1864b, 347) put it. God made "this World full of Beauty and Glory, every thing very good". But, laments Thomas Watson (1692, 67), sin "hath much eclipsed the beauty, sowrd the sweetness, and marred the harmony of the World".

A striking feature of nonconformist discourse is the reciprocity of mourning for the fallen world. Carol Adams (1995, 64), quoting Mickie Gustafson, notes that a "dead animal is more than just a dead body", if she is a loved animal whose life is grievable. It seems that a dead animal is more than just a dead body to the God who cares for them, as Jesus of Nazareth intimated in his reference to God's care for sparrows (Matt. 19.29). God, says Charnock (1864a:1681, 378; 1864b, 347), "was grieved at the heart" on account of His fallen, groaning creatures. "It is the joy of God to see all his works in their due order, every one pointing to their true end, marching together in their excellency, according to his first intendment in their creation". "The creatures were all created for the glory of God, and the support of man in the performance of his duty, who was obliged to use them for the honour of him that created them both. Had man been true to his obligations, and used the creatures for that end, to which they were dedicated by the Creator, as God would then have rejoiced in his works, so his works would have rejoiced in the honour of answering so excellent an end".

A feature of the dissonance sounded by the Fall was the loss of trust between humans and other animals. Creation could no longer rely upon humans to lead it in worship. Nonconformists wrote about creation's longing to praise God, and its groaning under a burden of suffering (Rom. 8). Men "that are not just to the brute-creatures", says Matthew Henry, "are furious and barbarous to them, [and] evidence, and confirm in themselves, a habit of barbarity, and help to make the creation groan". The groaning of creation, the dissonance of the song, does not go unheard (Sampson 2018, 91f). As Wesley (1872:1771, para 2.III.1) puts it, "while 'the whole creation groaneth together,' (whether men attend or not,) their groans are not dispersed in idle air, but enter into the ears of Him that made them". The creatures, warned Thomas Manton (1870:1684, 165), "will cry to God for revenge".

Egregiously, the very qualities of the creature which gave them their unique voice are appropriated by human power and transformed into groaning. Each animal, says S. G. De Graaf (1977, 34), "was given a nature of its own and a distinct purpose for its existence. In each of them a distinct thought of God is revealed. We do not know which animal to admire most - the lion for its majesty, the bull for its power, the horse for its spirit, the deer for its swiftness". These attributes have each been transformed into groaning: we shoot the lion for a majestic trophy, we fight the bull to show our superior power, we break the spirit of the horse, and we hunt the swift deer. A "plaint of guiltless hurt doth pierce the sky", says Philip Sidney (1999:1580, 225), and humans can claim: "we did this". The very human ability to conduct the chorus of praise has lead to tyranny; humans, says Andrew Linzey (1987, 98), "must exercise power [over animals], but only towards God's end".

"More righteous than I"

The whole world was created for this one purpose: to worship and glorify God its creator in one single chorus of song. Humans were created to conduct, but have conspicuously failed in their duty; human abuse mars the song, and sounds a dissonance among the players. The whole creation groans, yet it continues to sing as best it can, despite its conductor's sin. In fact, the Fall created a hierarchy in which humans were towards the bottom; the opposite to that found in Thomism. When we see an animal suffer, says Burroughes (1643, vol. 2, 67–8), "let us reflect upon our own hearts, and say, My sin is cause that this creature feels that evil that it doth, and say as Judah said unto Tamar, *She is more righteous than I*. So do you say... the creature is more righteous than I".[3] As Watson (1692, 9) observed, "Creatures below us, and above us, bring glory to God; and do we think to sit rent free? Shall everything glorify God but man? It would be a pity then that man was ever made". The notion that, in some circumstances, the interests of animals might outweigh those of humans was entirely alien to their times. Indeed, it is not common in ours, where it is most familiar in the eco-centrism of the deep ecology movement. Nonconformist discourse rejects eco-centrism, as it rejects anthropocentrism; but it does open the possibility that the interests of animals might have priority over those of humans, not for eco-centric reasons, nor to sustain them as a resource for human use, but because creation fulfils its meaning in song, not groaning.

Nonconformists also wrote of creation's longing for the revelation of God's children, and of restoration. As Stephen Charnock (1864b, 348) put it "The 'last time' is called not a time of destruction, but a "time of

[3] Burroughes here intertexts Gen. 38.26.

restitution and that of all things, (Acts iii.21)" when "all the fierceness of one creature against another will vanish. The world shall be nothing but a universal smile. Nature shall put on triumphant vestments".[4] Wesley (1872:1771, 2.III.2) cites Revelation 21.4: "the following blessing shall take place (not only on the children of men; there is no such restriction in the text; but) on every creature according to its capacity: 'God shall wipe away all tears from their eyes. And there shall be no more death, neither sorrow, nor crying. Neither shall there be any more pain: For the former things are passed away'". They looked not just to progress, but to redemption.

Disrupting the Order of Things

Creation is song sung for another, and humans are to conduct it to the glory of God and "the beauty of the whole" (Charnock 1864b, 23). This discourse destabilises both the anthropocentric hierarchy of dominant discourse, and subordinate discourse's reduction of animals to self-sufficient living subjects. It steps beyond the position of "the way things are", the place where each animal is located in a "global Order of Things", to the place of the *creature*, the created animal (Žižek 2008, 111, 115–6). In doing so, it opened a space for a novel ethic towards animals to consolidate. One that shifted the emphasis from a disorderliness in the Chain of Being, or a lack of rationality, to the particular animals themselves, and the human duty to enable all creatures to "bring glory to God". This entails the remarkable possibility that, not only do animals have covenant rights, but humans might have to admit with Burroughes "the creature is more righteous than I". It opens the possibility for an alternative imaginary which celebrates creation, especially animals, as lyric, as meaning, as harmony. Lynn White (1967, 1206) famously observed that Christianity has destroyed the sacred grove: "To a Christian a tree can be no more than a physical fact". But nonconformist discourse, rejects such self-sufficient existence; everything points to God, everything is sacred: "If the fulness of the earth is the Lord's, there is nothing in the world that is not sacred and pure", argues Calvin (1848:1546, vol. 1, 344). An immediate consequence of this is that unity

[4] Charnock's view of the "last time' was complex and nuanced, but beyond the scope of this discussion.

is no longer vested in human *being*, but in creations' shared task. For Martin Heidegger (1995, 176f, 186f), minerals are world-less, animals "poor in world"; only humans constitute a world. For nonconformists, all creation is world-rich in meaning, manifesting its creator.

Yet nonconformists were not blind to the cruelty they saw all around them. They heard the dissonances, even as they insisted that creation praises. Within their new imaginary, evil arises in a way that it could not in the Chain of Being discourse or in Enlightenment rationalism. In the former, predation, with its attendant cruelties, was permitted when the more perfect fed upon the less perfect; in the latter, predation is a regrettable consequence of Darwinism. In nonconformist discourse predation and cruelty are always, already harmony destroying, always the fruit of sin. Humans have marred the harmony of creation, and it groans; but it also looks towards restoration (Rom. 8.20f). The question arises: how to conduct a groaning choir in the knowledge that something is deeply amiss? The next chapter addresses this question.

References

Abbot, George. 1613. *An exposition upon the prophet Jonah*. London: Richard Field.

Adams, Carol. 1995. Woman-battering and harm to animals. In *Animals and women*, ed. Carol J. Adams and Josephine Donovan, 55–84. London: Duke Unversity Press.

Adams, Thomas. 1862:1629. Meditation upon some part of the creed. In The works of Thomas Adams, *vol.* 3, 85–264. Edinburgh: James Nichol.

Andrewes, Lancelot. 1650. *The pattern of catechistical doctrine at large*. London: Roger Norton.

Arrowsmith, John. 1645. *Englands Ebenezer*. London: Printed by Robert Leyburn for Samuel Man.

Arrowsmith, John. 1659. *A chain of principles*. Cambridge: John Field.

Bauckham, Richard. 2008. Creation's praise of god in the book of revelation. *Biblical Theology Bulletin* 38 (2): 55–63.

Bauckham, Richard. 2011. *Living with other creatures*. Waco: Baylor University Press.

Baxter, Richard. 1670. *The life of faith in three parts....* London: Nevil Simmons.

Berkouwer, G. C. 1962. *Man: The image of god*. London: Eerdmans.

Bouwsma, William J. 1988. *John Calvin*. Oxford: OUP.

Bradstreet, Anne. 1962:1650. Contemplations. In The works of Anne Bradstreet, 370–381. Gloucester Mass: Peter Smith.

Bridges, Charles. 1865. *An exposition of the book of proverbs*. New York: Robert Carter.
Bunyan, John. 1980:1686. A book for boys and girls. In The miscellaneous works, vol. 6: 183–269. Oxford: Clarendon.
Burdett, Sarah. 1839. *The rights of animals*. London: John Mortimer.
Burroughes, Jeremiah. 1643. *An exposition of the Prophesie of Hosea*. London: R. Dawlman.
Calvin, John. 1583. *Sermons on Deuteronomy*. Trans. Arthur Golding. London: John Harrison.
Calvin, John. 1845:1559. The Institutes of the Christian Religion. Macdonald.
Calvin, John. 1848:1546. Commentary on Corinthians. Edinburgh: CTS.
Calvin, John. 1855:1551. Commentary on first John. In Commentaries on the Catholic epistles. Edinburgh: CTS.
Calvin, John. 1958. Ethics and the common life. In *Commentaries, trans. And ed. Joseph Haroutunain*, vol. 23, 313–361. Philadelphia: The Library of Christian Classics.
Calvin, John. 1965:1554. A commentary on genesis. London: Banner of Truth.
Caryl, Joseph. 1670. *An exposition with practical observations upon the book of job*. Vol. 4. London: Thomas Sawbridge.
Charnock, Stephen. 1864a. A discourse on divine Providence. In *The complete works of Stephen Charnock*, vol. 1, 6–122. Edinburgh: James Nichol.
Charnock, Stephen. 1864b. Discourse on the existence and attributes of god, continued. In *The complete works of Stephen Charnock*, vol. 2, 8–554. Edinburgh: James Nicol.
Clarke, Adam. 1832. *Commentary on the bible*. New York: G. Lane and P. P. Sandford.
Clifford, Richard J. 1999. *Proverbs*. Louisville: John Knox Press.
De Graaf, S. G. 1977. *Promise and deliverance* Presbyterian and reformed.
Derrida, Jacques. 2008. *The animal that therefore I am*. New York: Fordham University Press.
Dod, John, and Robert Cleaver. 1612. *A Plaine and familiar exposition of the eleventh and twelfth chapter of the proverbs*. London: Thomas Man.
Doniger, Wendy. 1999. Reflection. In *The lives of animals*, ed. J. M. Coetzee, 93–106. NJ: Princeton University Press.
Hall, Joseph. 1654. *Select thoughts*. London: Nath. Brooke.
Heidegger, Martin. 1995. *The fundamental concepts of metaphysics*. Bloomington: Indiana University Press.
Henry, Matthew. 2008:1710. Commentary on the whole bible. Peabody, MA: Hendickson, Prov. 12.10.
Jamieson, Robert, Andrew Fausset, and David Brown. 1871. *A commentary on the old and new testaments*. Hartford: S.S. Scranton.

Keil, C. F., and F. Delitzsch. 1973. *Commentary on the old testament*. Vol. 9. London: Eerdmans.

King, John. 1864:1594. Lectures upon Jonah. Edinburgh: James Nichol.

Laland, Kevin N., and Bennett G. Galef, eds. 2009. *The question of animal culture*. Cambridge, Massachusetts: Harvard University Press.

Laplanche, Jean, and Jean-Bertrand Pontalis. 1980. *The language of psychoanalysis*. London: Hogarth Press.

Levinas, Emmanuel. 1969. Totality and infinity. Trans. Alphonso Lingis. Pittsburgh: Duquesne University Press.

Linzey, Andrew. 1987. *Christianity and the rights of animals*. New York: Crossroad.

Lipari, Lisbeth. 2009. Listening otherwise: The voice of ethics. *Int. J. of Listening* 23 (1): 44–59.

Luther, Martin. 1958. In *Works*, ed. Jaroslav Pelikan. Saint Louis: Concordia.

MacLaren, Alexander. 1908. *Expositions of holy scriptures New York a C Armstrong*. Vol. 17.

Manton, Thomas. 1870:1684. The complete works of Thomas Manton. London: James Nisbet, vol. 12.

Maynard, John. 1688. *The beauty and order of the creation*. London: William Gearing.

Milton, John. 1667. *Paradise Lost*. London: Peter Parker.

Northcott, Michael. 2020. Covid 19, human ecology, and the ontological turn. In *Pandemic, ecology and theology*, ed. Alex Hamilton. London: Routledge.

Oster, Malcolm R. 1984. The 'Beame of Diuinity': Animal suffering in the early thought of Robert Boyle. *Br. J. for the History of Science* 22:151–180.

Owen, John. 1851:1657. The strength of faith. In *The Works of John Owen*, ed. W. H. Goold, vol. 9, 19–52. New York: Robert Carter.

Rawlinson, John. 1612. *Mercy to a beast*. Oxford: Joseph Barnes.

Ray, John. 1701. *The wisdom of god manifested in the works of creation*. London: William Innys.

Rizzo, Daniella. 2024. Animal glossolalia: A Pneumatological framework for animal theology. *Pneuma* 46:60–79.

Sampson, Philip J. 2010. Humans, animals, and others. In *Intersections in Christianity and critical theory*, ed. Casandra Falke. Palgrave Macmillan.

Sampson, Philip J. 2018. *Animal ethics and the nonconformist conscience*. Palgrave Macmillan.

Seidenberg, David Mevorach. 2019. Veganism and Covenantalism. In *Jewish veganism and vegetarianism*, ed. Jacob Ari Labendz and Shmuly Yanklowitz, 161–194. New Yrok: Suny Press.

Sidney, Philip. 1999:1580. *The Old Arcadia* Oxford: OUP.

Stone, Ken. 2018. *Reading the Hebrew bible with animal studies*. Stanford University Press.

Tillotson, John. 1700. *The remaining discourses, on the attributes of god*. London: R Barker.
Toplady, Augustus M. 1825:c1778. Speech delivered at the Queens Arms, Newgate Street, on the following question, "whether unnecessary cruelty to the brute creation, is not criminal?. In *Works,* vol 3., 459–466. London: Baynes.
Tyrrell, James. 1694. *Bibliotheca Politica*. London: R. Baldwin.
Ussher, James. 1653. *A body of Divinitie*. London: Thomas Downes and Geo. Badger.
Venning, Ralph. 1669. *Sin the plague of plagues*. London: John Hancock.
Walker, George. 1641. The history of the creation. In *God made visible in his works*. London: John Bartlet.
Watson, Thomas. 1692. *A body of practical divinity*. London: Thomas Parkurst.
Wesley, John. 1872:1771. The General Deliverance. In *Sermons on Several Occasions,* vol. 2, Sermon 60 London: Methodist Publishing House.
Westminster Assembly of Divines. 1647. *Shorter Catechism*. London: A. M.
Wevers, Richard F. 1992. Concordance to Calvin's Institutio, 1559. *Grand Rapids: Digamma Publishers: fruere, delectare, and utere.*
White, Lynn. 1967. The historical roots of our ecologic crisis. *Science* 155 (3767): 1203–1207.
Whiten, Andrew. 2021. The burgeoning reach of animal culture. *Science* 372 (6537):
Wright, Tom. 2011. *Revelation for everyone*. London: SPCK.
Yates, John. 1622. *A Modell of Divinitie catechistically composed*. London: Fulke Clifton.
Žižek, Slavoj. 2008. *The fragile absolute*. London: Verso.

CHAPTER 6

Cruelty

Introduction

Previous chapters have explored creation as song, and the distinctive task of humans to enable singing. Yet the nonconformists were not blind to the cruelty and suffering all around them, and asked: how to conduct a groaning choir in the knowledge that something is deeply amiss? This chapter addresses that question.

Secular thought reads both animal suffering and predation as normal consequences of the evolutionary struggle for survival. Some consider cruelties entailed in human pursuits to be a part of this Darwinian Nature. Animal advocates more commonly rue suffering as inevitable among "wild" animals, while considering it unacceptable in the case of animals under human aegis. In each case, unavoidable suffering is seen as a normal, perhaps regrettable, part of nature. Nonconformist discourse presents a radically different view.

Creation as harmonious song implies that all animal suffering and predation is unambiguously *abnormal*, the fruit of the Fall, whether in "wild" or domesticated animals. Human sin has caused a "great rupture in the unity and harmony of all things"; there is a "dissonance in the world; a discord in the whole of created existence" (Woodhouse 2011, 63). Nonconformists argue that humans must not collude in the cacophony of cruelty; they are not to cause suffering for "sport", and prey animals

P. J. Sampson, *Nonconformist Perspectives on Animals and Language*, The Palgrave Macmillan Animal Ethics Series,
https://doi.org/10.1007/978-3-032-12158-5_6

are to be protected from predators where possible. This view has proved prophetically apt in our world dominated by human cruelty. Tone deaf humans, brought up to cruelty, must learn kindness. Following prophetic biblical texts, they envisaged a restored creation as peaceable, where the lamb would be *with* the wolf, not *within* the wolf (Is. 11.6 & 65.25). Normative human behaviour is modelled upon the harmony of both the prelapsarian and restored world, not on the fallen, predatory one. Deliberate cruelty perversely reverses this by substituting suffering and death for peace and life.

This chapter explores the consequences of this way of speaking for the way we treat animals, concluding with a discussion of gendered narratives of violence.

From Praise to Screaming

As we have seen, nonconformists widely held that animals praise God according to their kind. But animals not only make a joyful noise; in a fallen world, they also writhe and scream. Many suppose that, in the past, few cared about this either because they doubted that animals, lacking a soul, possessed the ability to suffer; or because they simply considered animal suffering unimportant. Thus Dix Harwood (1928, 75) concluded that "humanitarianism… is not to be expected in the period before 1700". Peter Singer (1991, 206; 2011, 64) attributes such a lack of humanitarianism to the belief that humans are "the special creations of God, made in the divine image and set apart from the animals". He especially emphasises the role of Darwinism in subverting the divine image: "That humans and animals are utterly different kinds of beings was unquestioned for most of the course of Western civilisation. The basis for this assumption was undermined by Darwin's discovery of our origins and the associated decline in the credibility of the story of our divine creation in the image of God". He observes of Bentham's concern for animal suffering in 1789, that it is "interesting that that thought was at least possible before Darwin" (Singer 2009).

The lack of humanitarianism in the dominant discourse of animals is well established, especially before the nineteenth century. Keith Thomas (1984, 17f, 25) has documented that the Tudor and Stuart age commonly exhibited an "aggressively despotic" language of "mastery", "conquest", and "dominion", rather than kindness in discussing animals; the "subjugation of the natural world" rather than "harmony and family accord".

Indeed, this largely remains the case in the twenty-first century. However, the "thought" that animals suffer long predates Darwin, or even Bentham's concern of 1789. Indeed, Mary Midgely (1996, 16, 112–3) has argued that the recognition of animal sentience and capacity to suffer was probably necessary for the relationships underlying domestication. Likewise, it underlay blood sports. Chillingly, to enjoy cruelty to animals, we must believe that they really do suffer; thus bullbaiting has not been replaced by "bulldozer baiting." Moreover, as we have seen, the thought that animal suffering is an important ethical issue also long predated Darwin, and is found widely in nonconformist discourse from the sixteenth century on.

The Suffering Body

It is necessary to believe that animals suffer if we wish to be cruel, or wish to avoid cruelty. Neither *property* nor Cartesian *mechanisms* can suffer, so to speak of animals as such is to distance ourselves from their suffering. But to be cruel, we not only have to believe in suffering, we also need to do things to animal bodies. Suffering is something we can inflict on animals because they have bodies of flesh, like ours. David Seidenberg (2019, 164) emphasises that animals have a fundamental standing in Torah which uses the same Hebrew vocabulary for human and animal bodies; the same applies to the Greek of the New Testament when it speaks of animal and human bodies, including the body of Christ. In English, we might say both human and animal bodies are *flesh*. If we wish to distance ourselves from suffering, and if the language of property or mechanism is not available to us, we speak not of flesh and bodies, but of meat; not of cow, but of beef; not of pig, but of pork or bacon. We definitely do not give personal names to meat, beef, or pork.

To continue to use the language of bodies and flesh is to risk disturbing conventional identities and cultural expectations, even of inviting horror, especially at points of transition, at the border between *animal* and *meat*, between human and animal *flesh*.[1] I have already mentioned the farming family, dining on the leg of a pig they had reared and named Ophelia, who abandoned their meal when their young son asked whether they were eating her right or her left leg (Wilkie 2010, 155–6). For them, in

[1] Here and in what follows, I draw upon Julia Kristeva (1982).

that moment, meat had become flesh, and they were touched by a horror; that of being confronted by the sheer experience of Ophelia's suffering, bodily reality, much like their own. Ophelia had become grievable.

To avoid this we separate talk about animal bodies from the norms and rituals which govern our ordinary behaviour. We devise new rituals, including that of buying meat from a butcher who has slaughtered and dismembered on our behalf; of calling in a pest control agency to kill for us. Clear boundaries and ritualised expertise in killing and dismembering enable us to exclude the suffering involved. The integrity of the animal's body has been destroyed for us, and either hygienically disposed of, or vacuum packed and re-labelled. We speak of those by whom horror may arise as experts, each with their own rituals to distance us from suffering: butchers, vermin exterminators, or slaughterers. Above all, meat is not flesh like ours. When this ritual language breaks down, when the family roast becomes Ophelia, the sense of horror all too readily returns.

Nonconformist discourse emphasises that all living creatures, including humans, manifest and praise God, and do so as living, embodied creatures. We share flesh, bone, and blood.[2] They read in their bibles that, as Joseph Caryl (1652, 207) put it, God has "betrothed himself in mercies and loving kindness" to all the creatures of earth: the Lord "has compassion on all he had made" (Ps 145.9). And they understood that they were to image God. This language provided far fewer resources for inventing rituals of separation, for avoiding horror at this conversion of praise into suffering. It makes it harder to draw a clear boundary between an animal's body and our own. In consequence, nonconformist discourse spoke more often, and more passionately, than usual about cruelty. When humans inflict suffering on other animals, we steal the voice of praise from them; only screaming is left, and this manifests nothing of God, although it tells us much about ourselves. God "aborreth all crueltie", said John Calvin (1583, 775).

Nonconformists recognised that animals suffer, and protested against it (Sampson 2018). Harwood (1928, 64f) acknowledges this, but denies that it showed concern for the animals; rather, he says it resulted from "protestant moralists" who denied the "worldly pleasures" of blood sports and hunting. The nineteenth century historian Lord Macaulay (1849, 159) went further by suggesting that such moralists obtained a

[2] The apostle Paul makes clear that there are different *kinds* of flesh, yet *flesh* they all are; see 1 Cor. 15.39.

perverse pleasure from their opposition to cruelty: the "Puritan hated bear-baiting, not because it gave pain to the bear, but because it gave pleasure to the spectators. Indeed, he generally contrived to enjoy the double pleasure of tormenting both spectators and bear". Perverse pleasure is also attributed to nonconformist protest by H. L. Mencken (1925, 59, col.1) who quipped that "the Puritan is someone who has the haunting fear that someone, somewhere, may be happy". Belden Lane (2011, 98, 120-1) wryly observed that neither Macaulay nor Mencken seems able to appreciate the passionate moral sensitivity to animal cruelty of the Puritans, or indeed of nonconformists generally. The attribution of perverse pleasure to Puritans marks, as we shall see, a defensive projection against the experience of horror.

Puritans, as do the overwhelming majority of people, recognised cruelty when they saw it. A cry of pain is different in a fox than in a cow or sheep, just as the leap of a spring lamb differs from that of a meerkat. We hear a cry, just as we see the leap of joy, and know the difference. Unless in urgent need of therapy, most people can distinguish praise (chirping birds, lambs gambolling, the sleepy satisfied grunt of a pig) from groaning (screaming of lambs or pigs, howling of a dog); and probably always could. However, to recognise that animals suffer is not the same as preventing the suffering. An integral part of most blood sports, animal fights, and 'sport' hunting is that animals suffer and die. Stephen Vantassel (2009, 134, 137) recognises that foothold traps cause animals to "suffer shoulder dislocations, cuts, bruises, swelling, broken bones, tooth damage and 'wring off'", but asks how much value should we place on the significance of an "animal's foot swelling" when it will be killed by the trapper anyway.[3] It may even be that suffering is intentionally increased. Temple Grandin (1988, 9) noted that a typical comment from a slaughterhouse worker with sadistic tendencies is, "It is going to die in five minutes so it does not matter how I treat it".

That so few care sufficiently to do anything about animal cruelty is no doubt complex, differing from one context to another, and utilising various defences against the cognitive dissonance which indifference to suffering is liable to provoke. One such defence is the attribution of perverse pleasure to those who do care. A remarkable feature of nonconformist discourse is that it was far more concerned about animals than the

[3] A "wring off" is the twisting or gnawing off of a body part such as a foot by a desperate animal.

dominant culture of their day mandated, and passionately protested on their behalf. Most agreed with Calvin that God "aborreth all crueltie"; some even considered that God identified with the pain of abused animals (Lane 2011, 121). As a result, they met indifference and ridicule from wits such as Macaulay or Mencken, and sometimes overt hostility (Sampson, forthcoming).

We will now turn to the nonconformist protest against cruelty in more detail.

Creature of God, Not Instrument of Man

The dominant discourse of animals provided, and continues to provide, few resources for protesting against cruelty to animals. If animals are property, the owner may do as he wishes, whether they suffer or not. Some followers of Descartes even denied that an animal's groans were more than mechanical reflexes. Should this seem reprehensibly hard-hearted, appeal can be made to the "traditional argument that animal suffering is morally inferior to that of human suffering". This appeal suggests that it is "not always an evil", simplifying the problem of pain by removing "the issue of animal suffering from the debate" (Vantassel 2009, 57, 107, 175).

The nonconformists found in the biblical text many points of resistance to such rationalisations, not least its condemnation of cruelty as *wickedness* (e.g. Prov.12.10). As John Ray (1701, 59-61) noted, this strong censure would not make sense if animals were automata which could not suffer. "Wickedness" firmly locates animal suffering within the language of evil; indeed, as we shall see, many nonconformists regarded it as demonic.

Thomas Aquinas had found these same "prohibitions of cruelty to dumb animals" in "holy Scripture", but his Aristotelian *anthropocentrism* implied that the evil arose from harm or loss to his fellow-men, not from pain to the animal. By contrast, the *theocentrism* of nonconformist discourse implied that animals are of value for their own sake, and cruelty requires no instrumental purpose to be a sin. Animals are individuals, created to praise God, and humans are to care for them. Moreover, the suffering of animals was a result of the theological *Fall,* and therefore a human responsibility.

The Fall

At creation, wrote William Cowper (2013, 1785, VI.364-7), there was neither cruelty nor suffering: "fear as yet was not, nor cause for fear". The Fall changed this. Yet animals remained creatures of God, making cruelty towards them an offence against God, something nonconformists took very seriously. Calvin imagines a cruel man saying "Tush, I care not, for it is but a brute beast", and he answers "Yea, but it is a creature of God'". This discourse remained influential into the nineteenth century. Sarah Burdett (1839, vi), a pioneer of the RSPCA, argued that "cruelty to animals is a heinous crime and sinful offence against the Almighty who created them".

Just how "heinous and sinful" may be judged by John Maynard's (1688, 19–20) scrupulous avoidance of animal suffering: "the withholding of due and necessary food from the Beast, or over-burdening it beyond measure, is a dishonour offered to its Creator. And as these Beasts of service must not be abused, because they are the Creatures of God; so neither may we in sport torment the poor Creature that crawleth upon the Earth... to abuse in sport (though it be as mean a Creature as a Flie &c.) is to take the Name of God in vain; since these works of His, are such things, by which, as by a name, he hath made himself known unto us; the least Flie being an evidence of his Almighty Power". To condemn sporting with a fly seems to point to an excessive sensitivity, but Maynard is not alone; William Cowper (2013, 1785, VI.583-90) refused to count among his friends any who unnecessarily tread upon a worm.[4]

Demonic Images

As we have seen, nonconformist discourse understood humans created in the Image of God to imply shepherding dominion over animals. If God "betrothed himself in mercies and loving kindness" to all the creatures of earth, if as the pslamist says, He has "compassion on all he had made", then so must those who seek to image Him. If we do not, we fail to image God. God made animals and "preserves them and is good to them", wrote John Dod and Robert Cleaver (1612, 141 - my emphasis),

[4] Their concern may appear slightly less eccentric when we recall the then contemporary belief that the smallest creatures have the greatest sensibility to enable them to manipulate objects too tiny for larger animals to sense.

so we "shew our selves *unlike unto him*, and hurtful to his, if we offer abuse and wrong unto them".

The fastidious avoidance of cruelty which we find in Maynard and Cowper reflects the nonconformist view that it is genuinely wicked. Cruelty to animals, says Augustus Toplady (1825:1778, 460, 465-6) bluntly, "is truly and properly criminal". "How dare we then be destroyers of an animal's ease, which we ought to promote; or wantonly deprive them of that life, which we cannot restore?". Spurgeon (1873, 332), in remarkably immoderate language for a nineteenth-century Baptist preacher, regarded animal cruelty as not only criminal, not only "unlike God", but demonic. The abuse of animals demands justice, he wrote, and "an Act ought to be passed at once, or Mr. Justice Lynch might for once be invoked to give the demon his reward in an irregular Manner"; this from a man who was "averse to all capital punishment". Other nonconformists elaborated, calling it "devilish" or "fiendish" at a time when devils and fiends were taken with utmost seriousness (Sampson 2018, 88).

Animals remained creatures of God after the Fall, but they were not unaffected. Death entered the world. As David Attenborough's TV series *Life* demonstrated, animals go to extraordinary lengths to survive death, but ultimately it is this struggle which drives the Darwinian evolution of life itself. As one commentator on Attenborough's series remarked, "without his benign narration, *Life* would essentially be about things killing things". Nonconformist discourse has a radically different understanding of life and death.

The Fall had two major consequences: predation disrupted the harmony of creatures; and sin perverted the holy pleasure of human relationships with other animals.

Disrupted Harmony

According to John Milton (1667, IX.1000-1001, 781-84), at the Fall, the earth "trembling from her entrails", "felt the wound, and Nature from her seat, sighing through all her works, gave signs of woe that all was lost". The mosaic of individual creatures fractured. In particular, says Calvin (1953:1559, 11.6), the Fall produced "the cruelty of brutes, which prompts the stronger to seize and rend and devour with dreadful violence the weaker animals. There would certainly have been no discord among the creatures of God, if they had remained in their first and original condition. When they exercise cruelty towards each other, and the weak need to be protected against the strong, it is an evidence of the disorder which has sprung from the sinfulness of man". The Fall also produced human cruelty: by "our sin there is growne enmity between us and God's creatures", concludes Jeremiah Burroughes (1643, vol. 1, 576). "If any creature of God be at enmity with us and hurtful and pernicious to us", argues George Walker (1641, 160-61), "it is for our owne sins, because so wee have broken the peace, and are at enmity with God by our sinfull rebellion; therefore creatures… are become our enemies, and they by our transgression are made subject to vanity and corruption, under which they groane together with us". Augustus Toplady (1825, 1778, 465–6) was typically forthright: "the miseries and hardships under which the brute creation labours are primarily owing to the sin of man".

This should, said William Perkins (1606, Bk. 3, 589), "rather moove us to mourne, than to rejoyce". Thomas Draxe (1613, 26-7) emphasises that our sin should cause us to *grieve* for the suffering animal: "if the poor dumbe creature, (bird or beast) bee in any paine and miserie, let us not joy, nor sport ourselves in its paines and torments, but rather be sorry for it, and be greeved for our own sinnes, which the silly and sinlesse creature smarteth for". Animals as individuals have a grievable life.

Perverse Pleasures

Creation, said Luther, was meant for "the admiration of God and holy joy" (Linzey 1987, 13). Sin made it possible for humans to corrupt "holy joy" into perverse pleasures. We must not, warned Walker (1641, 26, 160–61), "abuse any of God's creatures… to satisfy our sinful desires and pleasures". To take "pleasure, delight, and rejoicing in their destruction… is a kind of scorne and contempt of the workmanship of God our creator", "corrupt, sinful and against pure nature". Do not, says Matthew Henry (1710, Gen. 9.1-7), "torment them for… pleasure". Nonconformist discourse considered it perverse to enjoy animal suffering, a thing not simply detestable, but also a direct offence against God. The offence was intensified as predation and suffering were consequences of the Fall. The puritan, Thomas Wilson (1614, 519), reproved "the cruelty of such as sport themselves in the mutual murdering of the creature; the death and destruction whereof, being a part of the curse for our sins, wee may not make our recreation; Gods curse may not be sported with".[5] Dod and Cleaver (1612, 141-2) ask whether "our sinnes in Adam brought such calamities upon them, and shall we add unto them by cruelty in our owne persons? Have our corruptions bin a cause of that fierceness that is in many of them one against another, and shall we solace our selves in seeing them execute it? God forbid".

This way of speaking proscribed the routine abuses to which farm and labouring animals had always been vulnerable. Not only Maynard and Cowper showed a sensitivity unusual in their day. Some men, wrote Flavel (1968:1671, 163), "are cruel and merciless towards [beasts]... How often have I seen them fainting under their loads, wrought off their legs, and turned out with galled backs into the fields or highways to shift for a little grass; many times have I heard and pitied them groaning under unreasonable burdens, and beaten by merciless drivers, till at last, by such cruel usage, they have been destroyed, and then cast into a ditch for dog's meat". Richard Sibbes (1630, 41-2) recalls "certain Commandments which the Jews called the hedges of the Law: as to fence men off from cruelty, hee commanded they should not take the dam with the young, nor seeth the Kid in the mothers milke, nor muzzle the mouth of the Oxe...". Henry (1710, Prov. 12.10) draws the practical consequences; an animal deserves care "because it is God's creature, and in conformity to Providence, which preserves man and beast. The beasts that are under our care must be provided for, must have convenient food and rest, must in no case be abused or tyrannised over".

Tyranny and Blood Sports

God, said Joseph Hall (1837, 156), "hath made us Lords, not Tyrants"; cruelty is treachery against our dominion duty. Henry (ibid) insists that humans "must not be barbarous and cruel to the inferior creatures. They must be lords, but not tyrants". Yet humans did tyrannise over other animals, most obviously for perverse pleasure in blood sports such as bullbaiting, cock-fighting, and 'sport' hunting.

These practices are socially complex: violent beating or baiting tenderises tough flesh; both excitement and wagering punctuate boredom, and cement community cohesion. Perhaps most importantly,

[5] He exempts hunting for food "yet with the least cruelty".

blood sports were considered essential for maintaining the martial ardour of the nation, and hardening the hearts of labouring men against the blood and suffering of battle. Indeed, blood sports became part of the social reproduction of the warrior masculinity, as I shall discuss further in the next chapter. Nonconformist opposition to cruelty challenged a popular pastime and threatened the food supply; but it also disturbed a significant means of reproducing fighting men. In the early nineteenth century parliamentary debates, banning bull baiting was seen as both unmanly and unpatriotic. Nevertheless, nonconformists remained outspoken.

Dod and Cleaver (1612: 141-2) condemned men who "make a sport of making them [animals] miserable; when it is a pleasure to put them to paine: when it is a pastime to behold their torment and tearing. This proceedeth not from a tender heart; this is not the work of righteousnesse". Men, said Robert Bolton (1634, 156), "bloodily minded towards harmless beasts, discover our natural propensity to cruelty", a quality he did not value, prized as it might be on the battlefield. "Cockfighting and bull-baiting" observed C. H. Spurgeon (1873, 334), "were not only detestable things as involving needless torturing of living things, but as corrupting, depraving, and preparing for eternal perdition all who delighted in them". Views as uncompromising as these were entirely outside the dominant discourse which considered animals as property which could be lawfully treated as the owner wished. Moreover, they threatened the security of the state.

Revenge and Judgement

Animals, tormented for perverse pleasures, may seek justified revenge for their suffering, as both fact and warning.

Dod and Cleaver (1612, 142) warn that an animal rebellion might be justified; if we "put them to paine" for our pleasure, then our cruelty "may justly returne upon our selves, their teeth, or hornes, or pawes assaulting us, or else Gods owne hand by some other means, revenging their wrongs" For Edward Topsell (1599, 194), animals, formerly peaceable, "now in revenge of their miseries which for our sakes they endure, they kill and spoile, and teare us in peeces where ever they meete us".

Even if abused animals have no revenge in this world, their cries for justice still ascend to God. The "creature groans against us", says Thomas Manton (1870:1684, 182) "because of the slavery we put them to, they groan for vengeance…". "God will condemn us for cruel and unkind folk, if we pity not the brute beasts" says Calvin (1583, 770). Nowadays, few people would be much disturbed at the thought of God condemning them, but nonconformists took it very seriously indeed. Joseph Hall (1837, 156) warns that "We little thinke it; but God shall call us to an account for the unkind and cruell usages of his poore mute creatures". George Walker (1641, 160-1) admonished that, by cruelty we "provoke God against us, according to the example of our own doings, and to mete out to us by the same measure as we have done to his creatures, the living works of his hands". He that "hears the young ravens, may hear them, for 'he is gracious'", cautioned Trapp (1660, 48), and Humphrey Primatt (1992:1776, 16) warned that we will have to give a "strict account" of our conduct towards animals on the day of judgement. William Cowper (2013, 1785, VI.442-3), a man who knew much of the terrors of judgement, wrote of animal cruelty:

> "… many a crime, deem'd innocent on earth,
> Is register'ed in Heav'n; and these no doubt,
> Have each their record, with a curse annex'd.
> Man may dismiss compassion from his heart,
> But God will never".

Nowadays, the prospect of having to give an account for our cruelty towards animals at the judgement seat of God holds few terrors, but for nonconformists it was something to be dreaded.

Mercy and Restoration

Suffering, Yearning, and Restoration

Animals were created to disclose God and to praise His glory. A suffering or fearful animal groans, and her capacity to do either of these things is compromised. Yet, suffering animals retain a longing for the freedom to praise which once they enjoyed; their song takes on a fearful yearning, at once a calling upon God and a rebuke to humans. As David Dickson (1659, 20) observed, "created things are subject to …the abuse of the vainest men, who abuse the Creature to all manner of sin, and prostitute it to their service, to the fulfilling of their vainest lusts". However, "the Creature is not subject to this vanity willingly, or of its natural propensity, or its natural use, to which it was created of God, …and they patiently endure it". At the restoration of all things, this yearning will be satisfied when, says Thomas Goodwin (1655, 38), "that creature that hath groaned under mans lusts, shall then be fully restored to the glorious liberty of the sons of God".[6] This, recapitulates, says Jeremiah Burroughes (1643, vol. 1, 578), the "kinde of excellency as [animals] had at the first in the Creation".

The contemporary poet, Maya Angelou (1983, 16–17), catches the sense of this yearning in suffering when she speaks of the "fearful trill" of a caged bird who sings and longs for freedom. She draws the image from a poem by Paul Laurence Dunbar (1993, 1899, 102). Dunbar, evidently familiar with evangelical religion,[7] wrote that the caged bird is singing "not a carol of joy or glee, But a prayer that he sends from his heart's deep core, But a plea, that upward to Heaven he flings". The nonconformist plaint and yearning for restoration echoes down the centuries to colour Angelou's poetry of our own time. Angelou subsequently recapitulated this discourse in 2002 in her autobiography *A Song Flung Up To Heaven*, again echoing Dunbar.

The Duty of Mercy

This distinctive nonconformist discourse rejected human passivity in the face of cruelty to animals, and required mercy, for animals live a grievable life. Thomas Wilson (1614, 586, Rom. 8.19f) wrote that: "seeing all creatures partake with us in our punishment, it should cause us to be mercifull unto them which need our mercy, and be in danger". James Ussher (1653, 248) argues that the beasts are entitled to rest on the sabbath, "that we may shew mercy, even to the beast". A sensibility requiring mercy to all creatures, was, according to Thomas (1984,152) "eccentrically tender-minded by the standards of the age".

For nonconformists, hard-heartedness is a spiritual not just a moral fault, just as pity and mercy are spiritual virtues. "God desired that men should be habituated to gentleness", wrote Matthew Poole (2007, 1669, 406), quoting Hebraist François Vatable. As the nineteenth-century

[6] Thomas Goodwin's complex understanding of "the world to come" is beyond the scope of this discussion.

[7] Dunbar speaks of having once been "devout", of his parson "earnestly preaching of grace", and of a "revival meeting" where the Lord's "good grace" was flowing (1993, 131, 156). This is the language of nonconformity.

Baptist preacher C. H. Spurgeon declared, the "man who truly loves his Maker becomes tender towards all the creatures his Lord has made"; "in gentleness and kindness our great Redeemer is our model". Conversely, the man who is cruel to animals shows his lack of godliness. John Calvin (1583, 877) preached that if "a man spare neither his horse, nor his ox, nor his ass, therein he betrayeth the wickedness of his nature".

Mercy, then, is not just compliance with the law, but a tender heartedness which ensured the care and well-being of animals. William Hinde (1641, 35-6) set out a list of minimum requirements for care, including timely and good feed, and necessary care "which is all the wages that the poore dumbe creature doth require of his master, for his worke and labour". Men must:

- moderate an animal's "labour and his toyle, his travell and his paines, proportioning his burthen to his strength, and his labour to his power and ability to perform it"
- must not "overcharge him for… profit, nor overtoyle and tyre him out for… pleasure, much less to spend his spirits, and breath, and life and all, to serve… lust and vaine humour"
- must not "be fierce and furious, cruel and violent, in immoderate striking and beating, spurring, spurgalling, scutching and lashing, whipping and scourging…".[8]

This thorough cataloging of duties was intended to leave his parishioners in no doubt how seriously Hinde took animal suffering. In a small community, we may imagine that the parson's opinion carried weight. When Calvin (1583, 770) warned his Genevan farming congregation "God will condemne us for cruell and unkinde folke, if we pitie not the brute beasts", we may suppose that animal welfare standards around Geneva rose.

This standard of care must often have been inconvenient to subsistence farmers; it was certainly counter to the dominant discourse of the day. But the godly man went further, and also took care for the emotional wellbeing of animals, that they should be in good heart and know enjoyment as well as having enough to eat and drink. Thus Hinde adds that it is "a note and token of a godly and good man, to have a tender care and

[8] "Spurgalling" is to wound with a spur; "scutching" is to slap or scrape.

due regard of his beast, and that not only in preserving him in life to have his being, but in keeping him in heart and good liking, to have (in his kinde) a wellbeing also…". Nor was he alone in this. Writing in 1612, Dod and Cleaver (1612, 140–1) noted that the "love and kindness and good dealing [of a godly man]… flows over also to the cattle, and creatures which he.. has opportunity to preserve from hurt or perishing".

Nonconformist discourse is often closely reasoned but is far from unemotional. Denunciations of cruelty were passionate. Indeed, nonconformists insisted that Christian faith is "experimental"; it must be experienced.

Richard Baxter (1817:1650, 96) declared that he is the "best Christian… who has the readiest passage" from the brain to the heart. The christian man should be kind-hearted: "accostome your selves in such wise to bee so kinde hearted, and to doe no deede of crueltie, as that even the little birdes may feele it", preached Calvin (1583, 775). The reasoning of the "brain" must align with the compassionate feeling of the "heart". Cruelty, says Sarah Burdett (1839, 43), distresses those with "just and right feeling"; conversely, for Charles Bridges (1865, 121f, Prov. 12.10) the wicked lack "right feeling". Animal cruelty, says Calvin (1849, 1565, Ez. 4.14) bluntly, "ought to be detestable to men". It demands advocacy. If we "cannot open the mouth of the dumb,… we may and must open our mouth for the dumb", says Henry (1710, Num. 22.22–23).

The examples of cruelty given above involved mainly men. Similarly, the activities of the newly industrialising society from the eighteenth century, and those considered in the remaining chapters of this book, are pursued largely by men. Cruelty is gendered. It is to that aspect of nonconformist discourse in a changing world that we now turn.

Gender

The Language of Cruelty

The nonconformist sensitivity to animal suffering was entirely foreign to the dominant discourse which Thomas (1984, 21-2) summarises as: man has a god-given "charter of dominion" which justified killing animals for pleasure, including bear-baiting and cock-fighting; we owe them no duties. Hard-heartedness was considered a manly virtue, and compassion a "pitiful philosophy". Thomas here draws attention to the gendered language of animal cruelty which expresses and reproduces masculinity in daily life as the next generation becomes men (Bourdieu 2001).

Before the nineteenth century, boys were notorious for cruel games including torturing dogs, cats, pigs, birds, and reptiles (Thomas 1984, 147–8). Blood sports such as bullbaiting discouraged "softness" among young men; in particular, it prepared them for war. Animal abuse provided the means for reproducing warrior masculinity, and was defended using the very language of such masculinity. Thus bullbaiting was said to lend "an athletic vigorous tone to the character", and to promote a "martial spirit"; its abolition was opposed by William Windham, Secretary at War, a man known for his opposition to nonconformist opinion (Woodfall 1800, vol. 2, 299ff). Conversely, sensitivity to suffering was considered a feminine trait, a "pitiful philosophy". Sally Mitchell (2004, 327) argues that, by 1889, anti-vivisection "was increasingly viewed as a silly cause backed only by sentimental women". Nor was the gendered language of the dominant discourse restricted to the nineteenth century. As we have seen, Peter Singer (1975, ix) argues that "the portrayal of those who protest against cruelty to animals as sentimental, emotional 'animal-lovers' has had the effect of excluding the entire issue of our treatment of non-humans from serious political and moral discussion".

Nonconformist discourse subverted this gendered language which associated military ardour and reason with men, but "softness" and emotion with women.

Opposition and Ridicule

From the beginning, subversive nonconformist discourse met opposition, especially through language denigrating qualities associated with women. Nicholas Cox (1686, 2) was concerned that compassion for animals would breed "softness". In 1610, Sir William Cornwallis called compassion towards animals "womanish" adding, "Alas, this gentleness of Nature is a plain weakness" (Fudge 2006, 105). In the nineteenth century, men proposing legislation on bullbaiting and vivisection were said to have been beseeched to it by their wives; in the U.S., anti-vivisectionism was called "effeminacy" (Woodfall 1802, vol. 3, 295f). In due course, gendered denigration became misogyny. Charles Darwin wrote to *The Times* complaining of women who "from the tenderness of their hearts... and their profound ignorance" opposed all animal experimentation (Preece 2003, 416). In 1904, the *British Medical Journal* printed an anecdote about Darwin which gives the flavour of the times. It alleged that he had referred to a male opponent of vivisection as "a kind of *female Miss Cobbe*", managing to impugn both the masculinity of the man, and the femininity of the anti-vivisectionist, Frances Cobbe who had recently died (Mitchell 2004, 327).

Modern Cruelties

The nonconformist discourse of animals accompanied the growing cruelties of an industrialising and urbanising society. From the eighteenth century, the mass transport of animals, the centralisation of slaughter, the extensive use of dogs and horses for transport in towns, together with the growth of vivisection, all opened new vistas of cruelty, mainly pursued by men. By the mid-nineteenth century, 220,000 head of cattle and 1,500,000 sheep annually were "violently forced" into Smithfield market through the streets of London. "The cruelties... practised habitually on the wretched animals defies exaggeration... eyes deliberately knocked out, horns broken off, tongues torn..." (Anon 1849, 142). The "barbarous treatment of animals, the very sight of which renders the streets of our metropolis so distressing to every feeling mind" inspired William Hogarth (1833, 65) to engrave his celebrated *Four Stages of Cruelty* prints; Hogarth portrayed mainly boys and men.

Charles Bridges (1865, 121f, Prov. 12.10) appealed to the familiar nonconformist discourse in this new situation: "A righteous man regardeth the life of his beast, duly attends to its comfort, and never presses it beyond its strength. The brutal habits therefore, the coarse words, inhuman blows, and hard tyranny on the public roads is disgraceful to our nature. ... But why is this humanity marked as the feature of a righteous man? Because it is the image of our heavenly Father, who spreads his cherishing wings over his whole creation.... Must not then his children reflect his whole image of love? And is not the want of any feature of this image a mark of doubtful relationship to him?"

Along with cruelties on the streets, the horrors of vivisection grew from the late eighteenth century and were condemned by evangelicals such as Shaftesbury who wanted complete prohibition, warning against regulation which would only lull many into "a belief that by its provisions, protection was afforded". He described in Parliament the vivisection of a "very clever, lively young female dog" who endured, over a period of 3 months, the progressive destruction of her brain through two holes drilled in her head: "And that was the use they made of the creatures committed to their charge! that the account they would render of their stewardship! All he could say was—and he said it truly and conscientiously—that, in every respect, he would infinitely rather be the dog than be the Professor.... On what authority of Scripture... did they rest their right to subject God's creatures to such unspeakable sufferings?... The animals were His creatures as much as we were His creatures; and 'His tender mercies', so the Bible told us, 'were over all His works'" (Hansard 1879, vol. 248, 420f).

According to Frances Power Cobbe (1904, 648), the evangelical preacher C H Spurgeon wrote an open letter to the Anti-vivisection Society, but it was not made public because "the extreme strength of the *expletives* was considered to transgress the borders of expediency!".

The Subjugation of Nonconformity

The nineteenth century saw a variety of initiatives, from the founding of societies to protect animals and birds, through the opposition to vivisection, to legislative reform. These were often framed within the above nonconformist discourse. Bull baiting and cock-fighting were banned in England in 1835 and, although pockets of baiting still exist, few now defend it. Sport hunting in the U.K. was banned in 2004. Some illegal hunting persists, and it is occasionally defended citing biblical texts, but this is uncommon.

So far, I have traced the structure which nonconformist discourse built for discussing both animals, and cruelty towards them. In the following chapters, attention turns to concrete examples of human relationships with other animals which are associated with extreme suffering. I will trace the influence, and ultimate subjugation, of nonconformist discourse in three areas associated with animal cruelty and extreme suffering: hunting, warfare, and the meat industry. In each, the now subjugated nonconformist discourse provides points of resistance; and a source of inspiration for an alternative way of singing. We begin with hunting, the subject of the next chapter.

References

Angelou, Maya. 1983. Caged bird. In *Shaker,* Why Don't You *Sing?* Random House, 16–17.

Anon. 1849. Smithfield cattle market. The Farmer's Magazine, vol. 19(2), series 2, 142–143, February 1849. London: Rogerson.
Baxter, Richard. 1817:1650. *The Saints' Everlasting Rest* Glasgow: Blackie.
Bolton, Robert. 1634. *Some Generall directions for a comfortable walking with god*. London: John Legatt.
Bourdieu, Pierre. 2001. *Masculine domination*. Cambridge: Polity Press.
Bridges, Charles. 1865. *An exposition of the book of proverbs*. New York: Robert Carter.
Burdett, Sarah. 1839. *The rights of animals*. London: John Mortimer.
Burroughes, Jeremiah. 1643. *An exposition of the Prophesie of Hosea*. London: R. Dawlman.
Calvin, John. 1583. *Sermons on Deuteronomy* Trans. Arthur Golding. London: John Harrison.
Calvin, John. 1849:1565. Commentary on Ezekiel Edinburgh: CTS.
Calvin, John.. 1953:1559. Commentary on the book of the prophet Isaiah, vol. 1. Grand Rapids: Eerdmans.
Caryl, Joseph. 1652. *An exposition upon the book of job, (chapters 11-14)*. London: J. Macock.
Cobbe, Frances Power. 1904. *Life of Frances power Cobbe*. London: Swan Sonnenschein.
Cowper, William. 2013:1785. The Task and Select Other Poems. Ed. James Sambrook. Lonodn: Routledge.
Cox, Nicholas. 1686. *The gentleman's recreation in four parts....* London: Jos. Phillips... and Hen. Rodes.
Dickson, David. 1659. *An exposition of all St. Pauls epistles....* London: Francis Eglesfield....
Dod, John, and Robert Cleaver. 1612. *A Plaine and familiar exposition of the eleventh and twelfth chapter of the proverbs*. London: Thomas Man.
Draxe, Thomas. 1613. *The Earnest of our inheritance*. London: George Norten.
Dunbar, Paul Laurence. 1993. Sympathy. In *The collected poetry of Paul Laurence Dunbar*. Charlottesville: University of Virginia Press.
Flavel, John. 1968:c1671. Husbandry Spritualised or the Heavenly Use of Earthly Things. In The Works. Edinburgh: Banner of Truth, vol. 5, 3–205.
Fudge, Erica. 2006. Two ethics: Killing animals in the past and present. In *Killing animals, 99–119. The animal studies group*. Illinois: Urbanan University Press.
Goodwin, Thomas. 1655. The world to come. London: Printed in Popes-head Alley, and in Westminster Hall.
Grandin, Temple. 1988. Commentary: Behavior of slaughter plant and auction employees toward the animals. *Anthrozoos, 1* 4:205–213.
Hall, Joseph. 1837. Of Balaam. In *The works*, vol. 1. Oxford: D A Talboys.
Hansard. 1879. House of lords debate. July 15, 1879.

Harwood, Dix. 1928. *Love for animals*. New York: Colombia University Press.
Henry, Matthew. 1710. *An exposition of the old and new testaments*. London: Nisbet.
Hinde, William. 1641. *A Faithfull remonstrance of the holy life and happy death of J*. Bruen London: Stephens and C. Meredith.
Hogarth, William. 1833. *Anecdotes of William Hogarth*. London: J.B. Nichols.
Kristeva, Julia. 1982. *Powers of Horror*. New York: Columbia University Press.
Lane, Belden C. 2011. *Ravished by Beauty*. Oxford: OUP.
Linzey, Andrew. 1987. *Christianity and the rights of animals*. New York: Crossroad.
Macaulay, Thomas Babington. 1849. *The History of England from the Accession of James the Second*. Vol. 1. Leipzig: Bernard Tauchnitz.
Manton, Thomas. 1870:1684. The Complete Works of Thomas Manton. London: James Nisbet, vol. 12, Sermon 29.
Maynard, John. 1688. *The beauty and order of the creation*…. London: William Gearing.
Mencken, H. L. 1925. Clinical notes. The American Mercury, Vol. 4 (13), 59, col.1.
Midgley, Mary. 1996. *Utopias, Dolphins and Computers*. London: Routledge.
Milton, John. 1667. *Paradise Lost*. London: Peter Parker.
Mitchell, Sally. 2004. Frances power Cobbe: Victorian feminist, Journalist, Reformer University of Virginia Press.
Perkins, William. 1606. *The whole treatise of the cases of conscience*. Cambridge: John Legat.
Poole, Matthew. 2007:1669. *The Works*, vol. 1, Gen. 9.4. Culpeper, Virginia: Master Poole Publishing.
Preece, Rod. 2003. Darwinism, Christianity, and the great vivisection debate. *Journal of the History of Ideas* 64 (3): 399–419.
Primatt, Humphrey. 1992:1776. The Duty of Mercy Fontwell, Sussex: Centaur Press.
Ray, John. 1701. *The wisdom of god manifested in the work of creation*. London: William Innys.
Sampson, Philip J. 2018. *Animal ethics and the nonconformist conscience*. Palgrave Macmillan.
Sampson, Philip J., forthcoming. Men and Mercy: animals, masculinities, and language in English public policy.
Seidenberg, David Mevorach. 2019. Veganism and Covenantalism. In *Jewish Veganism and Vegetarianism*, ed. Jacob Ari Labendz and Shmuly Yanklowitz, 161–194. New York: Suny Press.
Sibbes, Richard. 1630. *The bruised Reede and Smoaking flax*. London: R Dawlman.
Singer, Peter. 1975. *Animal liberation*. New York Review.

Singer, Peter. 1991. *Animal liberation*. London: Harper Collins.
Singer, Peter. 2009. Darwin and the Animals. A talk by Peter Singer at the Ideas Festival, Bristol. http://animalrightskorea.org/essays/peter-singer-darwin-and-the-animals.html Accessed June 6, 2014.
Singer, Peter. 2011. *Practical Ethics*. Cambridge University Press.
Spurgeon, C. H. 1873. A word for brutes against brutes. *The Sword and Trowel* 3:332–338.
Thomas, Keith. 1984. *Man and the natural world*. Penguin.
Toplady, Augustus M. 1825:1778. Speech delivered at the Queens Arms, Newgate Street, on the following question, "whether unnecessary cruelty to the brute creation, is not criminal? In *Works*, vol. 3, 459–466. London: Baynes.
Topsell, Edward. 1599. *Times lamentation: An exposition on the prophet Joel*. London: George Potter.
Trapp, John. 1660. Annotations upon the whole bible, vol. 3, Proverbs to Daniel. London: R W.
Ussher, James. 1653. *A body of Divinitie*. London: Thomas Downes and Geo. Badger.
Vantassel, Stephen. 2009. *Dominion over wildlife*. Eugene: Resource Publications.
Walker, George. 1641. The history of the creation. In *God made visible in his works*. London: John Bartlet.
Wilkie, Rhoda M. 2010. *Livestock/Deadstock*. Philadelphia: Temple University Press.
Wilson, Thomas. 1614. *A Commentarie upon the Most divine epistle of St Paul to the Romanes*. London: W Jaggard.
Woodfall, W. 1800. *The parliamentary register: House of commons 2 April 1800*. London: John Stockdale.
Woodfall, W. 1802. *The parliamentary register: House of commons may 24 1802*. London: John Stockdale.
Woodhouse, John. 2011. *Colossians and Philemon*. Ross-shire: Christian Focus Publications.

CHAPTER 7

Hunting

Introduction

The previous chapter set out the nonconformist hatred of cruelty. This chapter traces the consequences of that hatred in man's (and it is usually men) pursuit of "game", a term which nicely combines entertainment, sport, and flesh.

Hunting discourse has a long history, and has mainly been used by elite men to describe and justify their manly exploits. Part of the dominant discourse of animals, it has positioned itself as normal, part of the natural pattern of hunting for food, a healthy pleasure, and pass-time. Along the way, it has appropriated religious, gendered, and, since the nineteenth century, Darwinian language. For the dominant discourse, hunting marked a *continuity* between man and nature, between animal predation and human consumption of other animals. Although hunting has always had its opponents, systematic criticism had to await the nonconformist discourse of animals.

Today, most animal advocates consider hunting to mark a *discontinuity* between man and nature. Animal predation may be normal in the sense that it is a lamentable feature of "nature", and an inevitable consequences of the evolutionary struggle for survival. But human hunting, although it

P. J. Sampson, *Nonconformist Perspectives on Animals and Language*, The Palgrave Macmillan Animal Ethics Series, https://doi.org/10.1007/978-3-032-12158-5_7

may also have evolved, is now seldom necessary for food; sport hunting is considered ethically indefensible in a humanitarian culture.[1]

The nonconformist discourse of creation as harmonious song recognises a *continuity* between predation and hunting, in that both are unambiguously *abnormal*, finding their common origin in the Fall. Cruelty towards animals is wickedness; pleasure derived from cruelty is egregious wickedness. Normative human behaviour is modelled upon shepherd dominion, not hunting dominion; holy pleasure stems from the admiration of God's creation, not from another's pain and death. In particular, nonconformist discourse had little good to say of the fighting, masterful virtues associated with hunting; not did it consider that bringing up children, usually boys, to hunt was a good way to produce just and compassion men. Rather, drawing inspiration from the harmony of both the prelapsarian and restored world, it considered that humans are to protect animals from predators where they come under human aegis, not exploit predatory behaviour by hunting.

This chapter explores this nonconformist discourse of hunting which proved so influential. It concludes with a discussion of the appropriation and transformation of that discourse by a contemporary evangelicalism which claims a nonconformist heritage. I suggest that the dominant discourse of hunting not only permits great suffering, but also authorises and valorises gendered narratives of violence. Both these features have filtered into modern evangelicalism through its embrace of hunting spirituality.

We begin by identifying major traits of hunting dominion.

The Dominant Discourse of Hunting

Sport hunting, especially of larger mammals, has long been associated with elites; the poor caught small mammals and "vermin", but this was largely as food (Cartmill 1993, 60f). By the sixteenth century in England, George Gascoigne (1575) could write that only "the studies of Divinitie, & grave discourses" were "more comendable or necessary for al Noblemen & Gentlemen" than that of hunting "bicause it is both

[1] By "hunting" I mean "sport hunting"; although this is a modern term, I will apply it retrospectively to hunting traditions whose principal purpose is to provide pleasure for the hunters. I include trapping, netting, etc., as well as shooting, spearing and so on. I occasionally refer to subsistence hunting for food, but that is not the focus of this chapter.

profitable and godly", an "honest recreation" "which doth maintaine the body in helth, the mynd in honest meditations". Hunting, said Nicholas Cox (1686, 1) "is commendable not onely for *Kings,* Princes, and the Nobility, but likewise for private Gentlemen: And as it is a Noble and Healthy Pastime, so it is a thing which hath been highly prized in all Ages".

Here, hunting is for elite men, with a strangely modern-sounding and pragmatic emphasis on physical and mental health. In so far as the Christian church has been aligned with social status and political power, it has often authorised hunting practices, usually deploying an appropriated "dominion" discourse which I shall term "hunting dominion".

Thomas Aquinas (1924, III.II.cxii) inferred from an Aristotelian *Chain of Being* argument that "the hunting of wild animals is just and natural, because man thereby exercises a natural right". This appropriation of dominion became part of the dominant discourse of hunting, and has persisted into modern times. In 1963 Gavin Maxwell described a Minister of the Church of Scotland who shot two otter cubs playing on the seashore. Maxwell protested, but the Minister reminded him that the Lord gave man dominion over animals (Bradley 1990, 13). Now, times have changed in the U.K.; hunting is largely banned, and criticism of "hunting dominion" has spread, including to some among the hunting elite. For example, before his coronation, King Charles III argued that the Genesis dominion narrative has been taken too literally, and "has provided Western man with "an overbearing and domineering attitude to God's creation", contributing "to a feeling that the world is somehow entirely man's to dispose of" (HRH 1991, 8); surely, a tacit rejection of hunting dominion. Nevertheless, the justification of "hunting dominion" persists among some secular elites in the U.K., and is almost ubiquitous in the evangelicalism of the U.S. This appropriation of dominion by the dominant hunting discourse enables us to trace the intersection of animal cruelty with masculinity and pleasure.

The Manly Man

From classical antiquity, the dominant discourse has pragmatically justified hunting as necessary to reproduce warrior masculinity: a man skilled in horsemanship, with military discipline, and inured to suffering. As Virgina Woolf (1943, 13) observed, "Scarcely a human being in the course of history has fallen to a woman's rifle; and the vast majority of the birds and beasts have been killed by men, not by us". In his description of a utopian society, Tommaso Campanella (1886:1602, 240–1) advises that boys prepare for manhood by exercises in hunting. Nicholas Cox (1686, 1–2—my emph.) recommended the "pleasures" of hunting as it "trains up youth to the use of ***manly*** exercises.., and does not encourage Softness". According to Baldesar Castiglione (1967:1528, 63—my emph.), hunting requires "a great deal of ***manly*** exertion". What animal baiting did for labouring men, hunting did for elites.

Killing the prey without modern weapons required both mastery and a heart hardened against pity; it cannot easily be done at a distance. The hunter becomes inured against the sight of "wounds and blood of slaughtered beasts", preparing the warrior for the bloody sights of the battlefield (Plutarch 1909, 959 c-e). Hunting taught young men to overcome any unmanly reluctance to thrust iron into living flesh and dismember the enemy.

Moreover, just as a labouring man is caught up in the excitement of the bull bait, so a manly ardour grips the hunter, preparing him for the frenzy of battle. This delirious spirit of the hunt survives into contemporary accounts: "the angry, eager feeling a huntsman ought to have that's out for blood, and means his hounds to get it", as Siegfried Sassoon (1928, 10) eloquently put it. If "the blood flows abundantly", says José Ortega, "it intoxicates, excites, maddens both man and beast, [since] blood has unequalled orgiastic power". Without this perverse ardour, "the spirit of the hunt disappears" (Cohn and Linzey 2009, 325).

The prey, pursued and conquered by a man in the grip of "orgiastic power", is correspondingly feminised. At best, this generates a discourse of perverse pity. Shakespeare (1623a, 3.1. 89–91) portrayed Lavinia, raped, hands cut off, tongue torn out, found "straying in the park, seeking to hide herself, as doth the deer, that hath received some unrecuring [incurable] wound". Similarly, he describes the abused Lucretia, debating whether to live or die "as the poor frightened deer, that stands at gaze, wildly determining which way to fly" (Shakespeare 1623b, 1149). The act of hunting draws in the prey despite herself, already a perverse subversion of subjectivity, to provide numerous misogynistic metaphors.

Lavinia's pursuit and rape is the "solemn hunting" of a doe: "we hunt not, we, with horse nor hound, but hope to pluck a dainty doe to ground". Lucretia's rape is described by no less than seven distinct hunter-prey metaphors. Moreover, hunting appropriates creation, and transforms nature into an opportunity for both violence against animals and against women by "brave boys". "The forest walks are wide and spacious", says Shakespeare (1623a, 2.1.112; 2.2.24–26; 2.1.114–131), "and many unfrequented plots there are fitted by kind for rape and villainy: single you thither then this dainty doe, and strike her home by force, if not by words:... The woods are ruthless, dreadful, deaf and dull; there speak and strike, brave boys, and take your turns; there serve your lusts, shadowed from heaven's eye, and revel in Lavinia's treasury".

This gendered language is not restricted to the sixteenth century. Marti Kheel (1999, 91–2) recounts a contemporary hunter safety training course. Bullets were "called 'balls', firing was called 'discharge', and when a bullet hit an animal it was called 'penetration'... If a bullet was accidentally fired before the intended moment, it was labelled a 'premature discharge'. The law of 'first blood'... [dictates that] whoever first 'penetrates an animal and draws the 'first blood' has the 'privilege' of 'finishing the animal off', and claiming the body of the animal as his own". It may lack the eloquence of Shakespeare but, as Kheel, notes, it has "a familiar ring".

Since the late nineteenth century, this gendered discourse has been appropriated by a more Darwinian one which makes hunting normative, while retaining traces of its origin. As Tom McIntyre put it: being "hunters is what made humans what they are... from millions of years of a hunting past that shaped everything from our bodies to our brains to our social relationships... And so the real aberration is not that some humans still hunt and kill, but that some do not" (Cartmill 1993, 230). By "humans", he means men. In the U.S., this is often framed by the folk tradition of conquering the frontier wilderness, civilising the violently transformed forest and wood described above (Franklin 1999, 30f). Hunting continues to reproduce masculinity as boys are initiated, but this is now transformed into demonstrating man's right to survive as the fittest, most vigorous predator, suited to a competitive society (Ibid 31f). It continues to inculcate what President Roosevelt called "fighting, masterful virtues" and "vigorous manliness", as boys become men through hunting; but manliness is transformed from dismembering the enemy into the symbolic performance of the hunt (Cartmill 1993,153). Blood continues to flow, but now for the mark of a boy's initiation ceremony rather than in the frenzy of battle. The heart is hardened, but now against excess empathy and sentimentalism rather than against the screams for mercy; indeed, some influential representatives of anthropocentric evangelicalism have recently argued against an "untethered" empathy which, it is said, might lead us to a misguided compassion for the suffering of another (Rigney 2025). The warrior masculinity of the battlefield has become metaphorical. Yet the blood, suffering, and misogyny remains.

Children

Some remnants of the earlier warrior discourse are lodged uncomfortably in their new Darwinian setting. For example, "blooding" children, the vestige of a more violent time, suggests a darker side to hunting, and an overlap with discourses of abuse. According to one U.S. hunter, some children have "to be chased down, tackled and held still to receive their initiations" (Dubuc nd). Under what other circumstances would it be acceptable to hold down children and smear blood on them from the mutilated body of a recently killed animal?

A Most Holy Cause

Hunting is not just an individual activity. As a means of reproducing warrior masculinity, it also has a collective, ceremonial aspect, a liturgical order of events which adumbrate the transcendent, the sacred; this locates suffering within a larger authorising frame, isolating the individual hunter from his cruelty. The grievous cruelties of the hunt require a transcendent language to authorise them; otherwise they might seem reprehensible. What better than the language of religion and spirituality. Erasmus (1876, 78–9) satirised this spiritual aspect of hunting in his description of the butchering of a deer:

> "When they have run down their game, what strange pleasure they take in cutting of it up. Cows and sheep may be slaughtered by common butchers, but what is killed in hunting must be broke up by none under a gentleman, who shall throw down his hat, fall devoutly on his knees, and drawing out a slashing hanger (for a common knife is not good enough), after several ceremonies shall dissect all the parts as artificially as the best skilled anatomist, while all that stand round shall look very intently, and seem to be mightily surprised with the novelty, though they have seen the same an hundred times before; and he that can but dip his finger, and taste of the blood, shall think his own bettered by it: and though the constant feeding on such diet does but assimilate them to the nature of those beasts they

eat of, yet they will swear that venison is meat for princes, and that their living upon it makes them as great as emperors."

Some four hundred years later, this sacramental aspect was noted by Henry Salt (1914) who called the initiation rite of "blooding" children a "gruesome parody of the rite of baptism".

From Plato (1988, 217) on, many hunters have confirmed this sacred aspect. Darwin (1999, 18) was a passionate hunter, calling it a "most holy cause"; he was so excited at killing a snipe that he had "much difficulty in reloading my gun from the trembling of my hands". Safari Club International members trophy-hunt magnificent wild species as a path to the sacred, "a spiritual act of love" (Scully 2002, 54f, 176f).

In sacred hunting, the manifestation and praise of God have been displaced by suffering and death. Cox (1686, 3), a keen advocate of gentlemanly hunting, recognised that there is a "great danger of becoming altogether addicted to Slaughter and Carnage". Calvin (1850, Jer. 9.5) puts it more bluntly. Giving hunting as an example of "men given to pleasure, when lust draws them here and there, who feel no concern for the greatest weariness"; "they were so devoted to wickedness, that the pleasure of doing evil wholly blinded them and made them mad". William Thompson, one of the founding fathers of the "sport" of bowhunting, said of shooting deer: "I have so loved them that I longed to kill them"—a sentiment which Cartmill (1993, 238) feels it reasonable to regard as pathological yet is, he disturbingly observes, "commonplace among serious hunters"; indeed, we just saw Scully say much the same of Safari Club International members.

For Calvin, this spiritual aspect of hunting is perverse, arising from "the pleasure of doing evil". However, as we shall see, modern evangelicalism has appropriated it, generating a new christian spirituality (Scully 2002).

Subordinate Discourses of Hunting

In Northern Europe, there has been a subordinate discourse alongside this dominant one, although very much in the minority.

Despite Thomas Aquinas' authorisation of hunting, the medieval church discouraged it among clergy, save for the military orders whose lion hunts maintained their warrior skills. Geoffrey Chaucer (1987, 1.1) could assume that his readers would recognise as a worldly man the Monk who "loved venerie". This carnal Monk is a "manly man" who followed new customs rather than the somewhat strict church tradition, and didn't give a fig for the "text" that says hunters are not holy men. Some influential but isolated individuals, from John of Salisbury to Erasmus of Rotterdam, were outspoken that Christians should avoid hunting. The Jewish tradition, drawing upon Torah, was consistently critical of sport hunting. In the eighteenth century, Isaac Lampronti judged that the "person who indulges in this sport is unworthy of the name of Jew" (Schochet 1984, 269).

More commonly, people recognised its cruelty without explicitly condemning it. Shakespeare's (1623c, 2.1.21) use of hunting imagery assumes that everyone will know that hunting is cruel. He laments that deer "in their own confines with forked heads have their round haunches gored"; and describes "a poor sequestered stag, that from the hunters aim had taken a hurt... The wretched animal heaved forth such groans that their discharge did stretch his leathern coat almost to bursting, and the big round tears coursed one another down his innocent nose in piteous chase". The "*piteous* chase [of] tears" is empathically related, but its mirror opposite of the hunter's *pitiless* chase of "hurt" is not explicitly condemned.

As I have noted above, a few not only recognised the cruelty of hunting, but also condemned it. In particular, Thomas More (1965:1516, 170–1; 151–2) anticipated many of the arguments later systematised in nonconformist discourse. He observed that hunting breeds a "cruel disposition... through the constant practice of such brutal pleasure", lamenting that "the hunter seeks nothing but pleasure from the killing and mangling of a poor animal". He has a rabbit being torn by dogs cry out to the huntsmen: "Insensate breed, more savage than any beast, to find cruel amusement in bitter slaughter" (Bradner and Lynch 1953, 149). Thomas More exhorted his readers "to be moved to pity when you see a poor little hare torn to bits by a dog, the weak slain by the stronger, the fearful by the fierce, the innocent by the cruel and merciless". However, such criticisms remained fragmented and very much subordinate. A more consistent critique only emerged from the sixteenth century within nonconformist discourse.

Nonconformist Discourse

It has long been recognised that nonconformists criticised hunting and baiting, but this has often been attributed to a dislike of worldly pursuits among "protestant moralists", rather than opposition to cruelty. Thus, William Cowper's opposition to hunting in the eighteenth century resulted, says Dix Harwood (1928, 64ff, 74–5, 320f), not from Calvinism, but from moralism and Enlightenment humanitarianism. This reduction of a complex nonconformist discourse to moralism has obscured its contribution to the wider debates about hunting.

Care for Wild Animals

The nonconformist condemnation of baiting applied principally to domestic animals, although some captive 'wild' animals were used, notably bears. However, hunting always concerns "wild" animals, those not under specific human aegis. The question for nonconformists, then, was whether God forbade cruelty to all animals, or just those domesticated for human use. This was easily answered.

Familiar with the biblical texts, nonconformists knew that God cares for wild beasts as well as domesticated ones. In the bible, wild animals are pictured as occupying the wilderness, the desert, the high mountains, and crags. Their care and provision is presented as the special province of God, who provides habitation, food, and drink for them without the intermediary of human agency (Job 12.10, 38–41; Ps. 104:10–27). Indeed, Jesus of Nazareth uses God's concern for wild birds to illustrate His fatherly care for humans (Matt. 10:29; Luke 12:6). Calvin (1583, 877, 775) asks why the Law of God forbids hunting wild birds when brooding (Deut. 22.6–7), and answers: "hereby hee meant to expresse the better, howe greatly hee abhoreth all crueltie. … [even though], to our seeming, there is no account to be made of [wild birds]".[2] "Abhoreth" is not too strong a word bearing in mind the promise to "prolong thy days" which is attached to the law protecting wild birds. The same promise accompanies the commandment to honour one's parents: "that thy days may be long upon the land" (Ex. 20.12, Deut 5.16). To care for a wild bird which, as

[2] Maimonides claims that this also protects fledglings in the nest, as they would not be worth taking (Seidenberg 2019, 177).

Calvin observes, seems of little account to most people, attracts the same divine blessing as honouring one's parents.

Tender Mercies

That God cares for all animals, domestic and wild, meant that the nonconformist discourse of cruelty could be carried over directly to hunting. Animals are God's creatures, created to manifest and praise Him as individuals, each unique, not as interchangeable quarry. Moreover, as with the cruelty of baiting, that of hunting was additionally reprehensible because the sole purpose of the sport was to gain pleasure by exploiting the consequences of sin. It is bad enough that human sin had made animal suffering unavoidable; to exploit such suffering for human pleasure is egregious wickedness. Dod and Cleaver (1612, 293) call hunting a "vaine sport", "slaverie to ... lust, and serving sin". John Gill (2012:1746–1763, Lev.17.13) says hunters are "more rustic and brutish" than others. All this was entirely incompatible with the dominant discourse of hunting. But a distinctive conflict arose over the role played by hunting in the reproduction of warrior masculinity.

Hunting dominion required hard-heartedness and violence; it is "a test of manhood" (Scully 2002, 55). In nonconformist discourse, ***normal*** manhood, corresponding to normal dominion, exhibited empathy, gentleness, pity, compassion, and mercy; not violence.

Philip Sidney (1999:1580, 225) rebuked hunters for their lack of empathy and compassion: "Thou fearest death; think they [hunted animals] are loath to die. A plaint of guiltless hurt doth pierce the sky". John Bunyan (1962, 15, 29) wrote of "a very great softness and tenderness of heart", nurtured by a sermon on the love of God. This is not a state of mind much valued in warrior masculinity, and can clearly not be reproduced through hunting. Hunting, say Dod and Cleaver (1635, 293), "maketh not the heart pitiful and mercifull; but it hardeneth the heart, and infecteth it with crueltie". Normal manliness is lost in sport hunting, and replaced by ***abnormal*** mastery and hardness of heart. By the nineteenth century, this discourse identifying manliness with tenderness was sufficiently well embedded in evangelical Christianity to provide an alternative masculine identity. For example, in a ballad by Mary Sewell (1858, 36), a young man is recommended for a job working with horses because he is gentle and kind to animals: "he felt himself a man", says Sewell, "although he did not flog his horses", the more common way of showing manliness. The norm of hegemonic masculinity enacted in hunting had been upended, and the dominant representations of both men and women could no longer be taken for granted. Neither could the reproduction of a perverse warrior masculinity through hunting.

Children

The Nonconformists rejected the dominant discourse's view that boys should be trained in manhood by hunting and cruelty. Their vision was just the opposite. In 1605, Lord Sheffield complained that his former Puritan tutor, Thomas Bywater, had held that hawks and hounds were "not ordained by God for man's recreation, but for adorning the world" (Lane 2011, 32).

The reforming archbishop, John Tillotson (1694, 90–1), advised his congregation to educate children in tenderness and pity which, he says "will be a good bar against Injustice and Oppression; and will be continually prompting us to Charity". He continues: "To preserve this goodness and tenderness of nature, keep Children out of the way of bloody Sights and Spectacles of cruelty; and do not allow them to torture and kill creatures for their sport and pleasure; because this will insensibly and by degrees harden their hearts, and make them less apt to be compassionate towards the wants of the poor, and the sufferings and afflictions of the miserable". This was a revolutionary view in its day,

Hunting entailed an abnormal and perverse relationship with animals, producing and reproducing an abnormal and perverse masculinity; it is a spiritual peril. It also generated perverse pleasures.

Pleasure

As we saw in Chapter 4, our first parent Adam took "unspeakable pleasure" in the creatures, nothing "could stain his purity of joy". We should, affirmed Calvin (1845:1559, 1.14.20), "take a pious delight in the clear and manifest works of God". This contrasts sharply with the perverse "brutal pleasure" of sport hunting; as Humberto Fontova puts it: "Hunters are simply guys who get a thrill out of killing animals" (Cartmill 1993, 229).

Where survival required killing a wild animal to eat, Matthew Henry (2008:1710, Lev. 1713) emphasised that the hunter must not gratify "the sensual appetite" but "ask questions for conscience's sake". Andrew Bonar (1966:1846, 326–7) similarly emphasises that the "chase was not to be an occasion for gratifying wild fiery passions; they were to hunt for food. ... It is not for the gay, wild spirits of youth; or, if fiery youth engage therein, it must lead them to the most solemn views of sin and righteousness". Animal suffering should lead us to solemn views, should cause us to mourn; to obtain pleasure from it is perverse.

William Cowper (2013:1785, vol. 2, 61) called hunting a "...Detested sport,

That owes its pleasures to another's pain;
That feeds upon the sobs and dying shrieks
Of harmless nature, dumb, but yet endued
With eloquence, that agonies inspire,
Of silent tears and heart distending sighs!
Vain tears, alas, and sighs that never find
A corresponding tone in jovial souls!".

Seeing Nature

The nonconformist discourse of hunting not only revised discourses of masculinity and pleasure, it also opened a space for an alternative way to see "nature".

The dominant discourse of hunting sees nature as a theatre of pursuit and pleasure. Cox (1686, 2) advises Gentlemen readers that "the Huntsman truely courts" the "innocent and natural delights... when he seeth the day breaking forth. When he heareth the chirping of small Birds pearching upon their dewy Boughs... How jolly is his Spirit, when he suffers it to be imported with the noise of Bugle-Horns, and the baying of Hounds, which leap up and play round about him".

But for nonconformists, creation is not a theatre of pursuit and pleasure, but a theatre of God's glory, a paean to God (Calvin 1845:1559, 1.14.20). The "'flying fowl'", says Bonar (1875, 41), "praise him ... 'turning all the air to music'" in earth's morning.

We saw earlier Shakespeare's hunting tropes transform wide and spacious forest walks into an opportunity for violence against both animals and against women by "brave boys". This is facilitated by a more general aesthetic of nature. Cox's advice on Gentlemen's recreations includes a celebration of nature, but, to anachronistically cite Tennyson, nature "red in tooth and claw". It is admirable, says Cox, "to observe the natural instinct of Enmity and Cunning, whereby one being as it were confederate with man … serves him in his designes upon others. … [H]ow soon will a *Hound* fix his Eye on the best and fattest *Buck* of the Herd, single him out, and follow him, and him onely, without changing, through a whole Herd of rascal Game, and leave him not till he kills him?".

Nonconformist discourse does not find this "admirable". Predation in nature is "evidence of the disorder which has sprung from the sinfulness of man", for which we should rather mourn than rejoice. Wherever animal relationships come under human aegis, we are to protect the weak against the strong. This divergence from the dominant discourse is exemplified in the biblical contrast of *hunter* and *shepherd*.

Biblical Hunters and Shepherds

Hunter Dominion

There is no royal hunting tradition in the bible, and Seidenberg (2019, 162–3) notes that Torah "virtually bans" it. Few nonconformists had anything positive to say about hunters. From Martin Luther to John Calvin in the sixteenth century, from Richard Baxter to Thomas Watson or John Flavel in the seventeenth, from John Wesley to John Gill in the eighteenth century, all regarded hunting as at best a waste of time, at worst as standing for devilish wickedness.

Philip Stubbes observed that there are many hunters found in the Bible, but none is *righteous*. (Sampson 2001, 82). For the most part, said Thomas Adams (1630, 116), hunting is "taken in the holy Scripture in the worst sense". Indeed the list of biblical hunters is formidable, including Cain, Nimrod, Esau, and the rider of the white horse (Rev. 6). Herod, who pursued the child "born King of the Jews" "to destroy him" (Matt. 2.2, 13), was a keen sport hunter.[3] None was a good role model for a faithful Israelite.

Nimrod is an emblematic figure, for he is the most famous human hunter in the bible: Nimrod "was a mighty hunter before the Lord" (Gen. 10.9). But, by the sixteenth century, it had long been known that this was not a commendation. John of Salisbury (1990:1159, book1.4) had concluded by 1159 that "all scholars condemn him". In fact, Nimrod

[3] Cain was reduced to hunting after the murder of his brother when he became a wanderer, not a farmer. For Herod, see Josephus (2010, 1, 21.13).

exemplifies the tyranny of hunting dominion: "tyranny, initiated by a huntsman to insult the Creator, finds its sole source in one who, amid the slaughter of beasts, wallowing in blood, learned to feel contempt for the Lord". He was "without feare of his Majestie [ie God]", says Adams (1630, 116). Calvin (1965:1578, vol. 1, 317) called him "the tyrant... branded with an eternal mark of infamy", "robust in hunting, or in violently seizing upon prey". The modern calvinist, Arthur Pink (1981, 7, 131), considered him be in brazen defiance of his Maker, a type of the Antichrist.

But the most prominent biblical hunter is the Devil who is renowned for setting traps, nets, and snares, and for digging pits for his prey. Satan, says Calvin "hunts his prey", and the devils, says William Gurnall (1669, 95), "will hunt thee to thy grave". Indeed, death itself is the exemplary hunter, consumer of all things, "fed and filled Immeasurably, all things shall be your prey", says Milton's Satan (Milton 1667, II.843–5). This association between the devil, death, and hunting echoes the demonic nature of cruelty itself in nonconformist discourse.

The perception of hunters as enemies of the Kingdom of God is exemplified by the nonconformist identification of the hunted animal of Psalm 22 with Jesus of Nazareth, and the hunters with his persecutors. John Flavel (1982, 149) tells us that the psalm "contains a most exact description of Christ's sufferings". The hunter "singles out [the deer] to hunt down that day;... 'Dogs have compassed me, the assembly of the wicked have enclosed me.' ... Thus was Christ hunted...". Spurgeon (2003, March 26) extends the soteriological typology: "'The hind of the morning' has drawn the cruel hunters upon himself, and now the most timid roes and hinds of the field may graze at perfect peace among the lilies of his loves".

Shepherd Dominion

It was the shepherd, not the hunter, that should point us to God. John Owen (1652, 418–9) contrasts Nimrod "the great hunter" with Christ "the great shepherd": the first founded the kingdom of Babel, the second the kingdom of Zion.[4] Indeed, the biblical list of shepherds is as distinguished as that of hunters is notorious: Abraham, Abel, Jacob,

[4] For Owen, "Zion" signified the Kingdom of God; it had no connection with contemporary Zionism.

Joseph, Rachel, Moses, David, Amos, Jesus and, according to Psalm 23 and elsewhere, God himself. This was no casual association. The cultural significance of the identity of *shepherd* is apparent in Joseph's brothers insisting that this was their occupation, contrary to Joseph's advice, and at risk of displeasing Pharaoh (Gen. 46.31–47.6). Ezekiel's good shepherd feeds and protects his flock, while the wicked shepherds hunt and eat them (Ez. 34). The shepherd *Rachel* bears the name meaning "ewe", pointing to the identification of good shepherds with their sheep. The good shepherd, says Jesus of Nazareth, has a kindly heart towards his sheep, is their sanctuary, and even gives his life for them (John 10). As Spurgeon (1915, 1) put it: "See that poor soul hunted by the dogs of hell. It flies swift as the wind, but faster far do the hunters pursue.... till Jesus Christ shall open his bosom for that poor hunted thing to hide itself within, it shall have no peace".

Modern Evangelicalism

Hunting discourse in the U.K. is now mostly nostalgic, and only occasionally appeals to biblical texts. In the U.S., sport hunting is declining, with only about 4% of the population in 2016 identifying as hunters, mostly from higher income groups. Yet it carries disproportionate symbolic and emotional weight there, especially within contemporary anthropocentric *evangelicalism*.[5] This latter is a contested term, and my usage parallels that of Robert Letham (1995, 12) who argues that *evangelicalism* is a departure from the nonconformist tradition. The latter, he says, is exemplified by the Westminster Confession, tacitly adopted by the eighteenth-century Evangelical Revival, which is corporately theocentric in affirming that all things, including humans, exist for the glory of God. By contrast, he sees the *evangelicalism* which emerged from the mid-nineteenth century as essentially anthropocentric, foregrounding individual human experience in conversion and sanctification. Despite this contrast, the new evangelicalism saw itself as the heir to nonconformity, especially to the Evangelicals of the eighteenth century. In the twentieth century, this perception of continuity with the nonconformist tradition remained, but evangelicalism itself further narrowed to focus upon *expressive conversionism*; and upon *biblicism*, a devotion to the bible (Bebbington 1989,

[5] See the discussion in the Preface for the distinction between "Evangelical" and "evangelicalism".

5–15). Hunting is part of the culture of modern American evangelicalism (Stoddard 2008).

The nonconformist discourse which was critical of hunting was taken to North America by Puritan settlers, and was commonplace there for more than a hundred years. Modern hunting discourses emerged only in the late eighteenth century, when the ability to shoot became a useful skill in the conflict with the English Crown. This was subsequently reinforced by the self-reliance ideology of the Frontier folk tradition, and by the Darwinism of the early twentieth century, to form a variant of the dominant discourse (Herman 2007). Modern evangelicalism presents a distinctive version of this discourse by appropriating nonconformist language, which is radically transformed in the process.

Spiritual Hunting

In nonconformist discourse, animals were created to manifest and praise God, and they do so individually, each unique and valued for its own voice; hunting is a spiritual peril. Evangelicalism appropriates this spiritual status, and transforms it from disdain for hunting into celebration of hunting. Animals become quarry, and hunting is embraced as a spiritual sport, rather than shunned as a spiritual peril. For evangelicalism, either God does not require humans to have compassion, or to care for animals individually; or else such "care" as He does require permits sport hunting with its associated suffering (Vantassel 2009, 56–60). Celebration of the spirit of hunting can carry an echo of the perverse intoxication characteristic of the dominant discourse described above. Thus Mike Huckabee (2007), former Southern Baptist minister and prospective Presidential candidate, described his hunting success to an NRA conference: "And somehow, by the grace of God, when I squeezed the trigger, my Weatherby 0.300 Mag … did its work, and somehow the angels took that bullet and went right to the antelope, and my hunt was over in a wonderful way". Huckabee explicitly narrated hunting as a spiritual blessing rather than peril; as celebrating angelic collusion in violence, rather than condemning demonic complicity.

A distinctive feature of evangelicalism's version of the dominant discourse is its affirmation that the world was created by God, with the opening chapters of Genesis read as the authoritative account. This is not a view widely held by the dominant discourse generally. However, the nonconformist narrative of those chapters of Genesis is appropriated and transformed. The vegetable diet of Eden vanishes into an authority to hunt; and the nonconformist wonder at creation becomes a perverse form of awe. In place of creation's manifestation of God, we find an appreciation of creation through violence. This enables Doug Phillips (2004) to say: "It is our mission to encourage such hunting, trapping, skinning and self-sufficiency skills to give our children a greater appreciation for the wonders of creation". This is a long way from Tillotson's advice to keep children "out of the way of bloody Sights and Spectacles of cruelty". Whereas nonconformist discourse widely recognised that the "dominion mandate" of Eden authorised a vegetable diet, evangelicalism reads it as though it were a *domination* mandate which endorses hunting. Thus, for evangelicalism, to reject hunting is to reject dominion; worse, it is to reject the bible. As Phillips observes, opposition to hunting "is a direct attack on Genesis and the dominion mandate".

Christian Bowhunter, Wayne Vining (2007), similarly writes within the hunting discourse which has appropriated the nonconformist Genesis narrative. He acknowledges that Eden's diet was vegetable, but says hunting "has been part of man's existence since [the] fall". For nonconformist discourse, only wicked men, if any, began hunting after the Fall; there was no grant to do so until the Flood narrative, and even then it was deprecated as undesirable. Vining continues by transforming nonconformist dominion into hunting dominion: "God gave us dominion over the animals, even telling us to 'kill and eat'". The latter clause is taken from Acts 10–11 which concerns the inclusion of the gentiles in the early Church, not the dominion of Eden. Vantassel (2009, 23) explicitly distances himself from nonconformist discourse by affirming that dominion is the exploitation and "conquest of nature", including hunting and trapping, not a "caring role of trust and leadership".

This transformation of Eden's dominion into hunting dominion entirely subverts the nonconformist resistance to the dominant discourse, and aligns itself with the contemporary Darwinian consensus. It has consequences in three areas: suffering and pleasure; consuming the prey; and raising children.

Suffering and Pleasure

In nonconformist discourse, humans have dominion as the conductor of creation's song of worship. Animal suffering is a consequence of sin and inhibits both creation's ability to praise, and the human ability to conduct the choir. In particular, nonconformist discourse rejects taking pleasure in hunting; we should not rejoice in the ravages of sin. Instead of violence and suffering, we find gentleness and awe.

By contrast, the dominant discourse of hunting "owes its pleasures to another's pain", as Cowper put it. This is said to be legitimate, either as a natural or an evolutionary right. Often peaceable animals like deer are hunted, but a higher status is attached to larger, more dangerous animals. Joseph Wood Krutch puts it bluntly, saying that the hunter's only satisfaction is to be able to say "something which wanted to live is dead" (Cartmill 1993, 228). The hunter has the power to make someone not-exist. The iconic image of the hunter, weapon in hand, victoriously standing with one foot on the kill, exemplifies this.

Modern evangelicalism embraces the pleasure motif of the dominant discourse. Reference to the sin which occasioned hunting is as rare in modern evangelicalism as it is common among nonconformists. This inevitably downplays suffering, subverting the nonconformist horror at the consequences of human sin. Some even ask whether animal suffering is a problem, and suggest that "humans may inflict and/or ignore a fair amount of animal suffering" when hunting (Vantassel 2009, 167; also 111, 172).

Consuming the Prey

In nonconformist discourse, animals must sometimes be hunted for food, as scarcity is a consequence of the Fall. When it is necessary, hunting is accompanied by mourning for sin, not pleasure in killing. Subsistence hunting is for nutrition, not pleasure. Thus, for Luther (1960, 132), Adam was the "gentle master to animals, rather than their slayer or consumer". As Thomas Adams (1862:1629, 137) observed "it was sin that made us butchers, and taught the master to eat the servant'. Luther's reference to "consuming" and Adams' to "butchers" reiterates the absence of flesh-eating in Eden, let alone of hunting.

For dominant discourse, including that of evangelicalism, the relationship between hunting and consuming the kill is reversed. Sport hunters hunt for pleasure, not food. The animal's carcass is eaten as a bonus to the pleasure of the kill. Perhaps under the influence of the Frontier folk narrative, or just as a response to criticism, the consumption of the carcass is often emphasised as though this justified the pleasure at killing. Where the prey species is not commonly eaten, this can provide bleak comic effect. In 2012, Daniel Richards, then responsible for the "management and wise use" of California's wildlife, attracted criticism when he shot a mountain lion while on a hunting trip in Idaho. He was reported as defending his decision, "saying he did not shoot the animal for pleasure and that he had eaten the lion" (Stevens 2012).

Evangelicalism's version of hunting dominion adopts and qualifies the dominant discourse by shifting the emphasis from pleasure to consumption, often advocating that the animal carcass should always be eaten, "out of respect" or to avoid waste. Some hunters have told me that they take pride in doing their own butchering for the freezer. Yet, as in the dominant form, it is the pleasure of the chase which motivates the hunt in the first place.

Raising Children

Evangelicalism naturally wishes to pass the spiritual blessing of hunting to the next generation, usually to boys. We have already seen Phillips mission to encourage hunting in order to give his children "a greater appreciation for the wonders of creation". James Dobson (2017) speaks warmly of hunting trips with his father, and Phillips writes that teaching a boy to hunt prepares him "to be a better man". But Phillips emphasises that this is a spiritual, not just an ethical matter. Every time "we train a boy to hunt for the glory of God and teach them the biblical implications of such hunting — we strike a blow against the Devil!". Nonconformists considered that the gentleness, pity, compassion, and mercy of normal masculinity cannot be reproduced through hunting and, as we have seen, discouraged it; hawks and hounds were "not ordained by God for man's recreation, but for adorning the world" as Lord Sheffield's Puritan tutor put it.

Hunting and the Kingdom

This chapter has set out the nonconformist vision of kindness and gentleness towards animals as a fruit of the Spirit, and their rejection of pleasure taken in suffering and death as wickedness (Gal. 5.22f). In the Kingdom, not even wolves will hunt (Is. 11, 65). If God has compassion on all he has made (Ps. 145.9), how can his servants do less? I have also discussed modern evangelicalism's appropriation of nonconformist *dominion*, and its transformation into conquest and exploitation. In doing so, "nature" has been transformed from the "theatre of God's glory" to a pleasure-park of pursuit. The new evangelicalism also reversed the nonconformist revolution in child care; from being shunned as a spiritual peril, hunting became a spiritual blessing to be passed on to the next generation. This is very far from anything which nonconformists would have recognised as their faith. Philip Stubbes (1877:1583, 182) was blunt: "Is he a Christian man, or rather a pseudo-Christian, that delights in blood?".

The next chapter takes up the norms of manly pursuits exhibited in hunting discourse, and traces their development into the masterful, fighting spirit required in warfare.

References

Adams, Thomas. 1862:1629. *The Workes of Thomas Adams*, vol. 3. London: Nisbet.

Adams, Thomas. 1630. *The Workes of Thomas Adams....* London: Thomas Harper for John Grismand.

Bebbington, D. W. 1989. *Evangelicalism in Modern Britain: A History from the 1730s to the 1980s*. Abingdon: Routledge.

Bonar, Horatius. 1875. *Earth's Morning*. New York: Robert Carter.

Bonar, Andrew. 1966:1846. *A Commentary on Leviticus*, Lev 17.13–14. Edinburgh: Banner of Truth.

Bradley, Ian. 1990. *God is Green*. London: Darton, Longman and Todd.

Bradner, L., and C. A. Lynch. 1953. *The Latin Epigrams of Thomas More*. University of Chicago Press.

Bunyan, John. 1962. *Grace Abounding to the Chief of Sinners*. Oxford: Clarendon.

Calvin, John. 1583. *Sermons on Deuteronomy*, trans. Arthur Golding. London: John Harrison.

Calvin, John. 1845:1559. *The Institutes of the Christian Religion*. Macdonald.

Calvin, John. 1850. *Commentary on Jeremia*h. Edinburgh: CTS.

Calvin, John. 1965:1578 *Genesis*. London: Banner of Truth.

Campanella, Tommaso. 1886:1602. The City of the Sun. In *Ideal Commonwealths*, ed. H. Morley. London: Routledge.

Cartmill, Matt. 1993. *A View to a Death in the Morning*. Harvard University Press.

Castiglione, Baldesar. 1967:1528. *Book of the Courtier*. Harmondsworth: Penguin.

Chaucer, Geoffrey. 1987. *Canterbury Tales*. Ed. Larry D. Benson, 165–178. London: Houghton-Mifflin.

Cohn, Priscilla N., and Andrew Linzey. 2009. Hunting as a Morally Suspect Activity. In *The Link between Animal Abuse and Human Violence*, ed. A. Linzey, 315–328. Sussex Academic Press.
Cowper, William. 2013:1785. The Task and Select Other Poems. Ed. James Sambrook, book III, 121–2, lines 326–329. London: Routledge.
Cox, Nicholas. 1686. *The Gentleman's Recreation in four parts....* London: Jos. Phillips... and Hen. Rodes.
Darwin, Charles. 1999. *Autobiography*. iRead Books.
Dobson, James. 2017. *A Great Father*. http://drjamesdobson.org/articles/fathers-family-faith/a-great-father. Accessed 6 May 2017.
Dod, John, and Robert Cleaver. 1612. *A Plaine and Familiar Exposition of the Eleventh and Twelfth Chapter of the Proverbs* London: Thomas Man.
Dod, John, and Robert Cleaver. 1635. *A Plain and Familiar Exposition of the Ten Commandments*. London: Thomas Man.
Dubuc, Don. n.d. The Blooding Rite—Why Do We Do It? http://www.dontheoutdoorsguy.com/features/featured-stories/51-the-blooding-rite-why-do-we-do-it.html. Accessed 27 Feb 2016.
Erasmus of Rotterdam. 1876. *In Praise of Folly*. London: Reeves & Turner.
Flavel, John. 1982. Sermon 19. Of Christ's Humiliation in his Life. In *The Works of John Flavel*, vol. I. Edinburgh: Banner of Truth.
Franklin, Adrian. 1999. *Animals and Modern Cultures*. London: Sage.
Gascoigne, George. 1575. *The Noble Arte of Venerie or Hunting....* London: Christopher Barker, unnumbered pages preceding text.
Gill, John. 2012:1746–1763. *Exposition of the Entire Bible*. Amazon reprint.
Gurnall, William. 1669. *The Christian in Complete Armour*. London: Ralph Smith.
Harwood, Dix. 1928. *Love for Animals*. New York: Colombia Univ Press.
Henry, Matthew. 2008:1710. *Commentary on the Whole Bible*. Peabody, MA: Hendickson.
Herman, D. J. 2007. From Farmers to Hunters. In *A Cultural History of Animals in the Age of Empire*, ed. K. Kete, 47–72. Berg.
HRH The Prince of Wales. 1991. Foreword. In *Save the Earth*, ed. Jonathan Porritt, 6–9. London: Dorling Kinderseley.
Huckabee, Mike. 2007. Remarks by Former Governor Mike Huckabee (R-AR) at NRA's "Celebration of American Values" Conference in Washington, DC, 21 September 2007. https://www.nraila.org/articles/20070921/remarks-by-former-governor-mike-huckabee-r-ar-at-nras-celebration-of-american-values-conference-in-washington-dc-92107.
John of Salisbury. 1990:1159. *Policraticus* Cambridge: CUP.
Josephus, Flavius. 2010. *The Wars of the* Jews. Ed. William Whiston. Digireads, Book 1.

Kheel, Marti. 1999. License to Kill. In *Animals and Women*, ed. Carol Adams and Josephine Donovan, 85–125. Durham: Duke University Press.
Lane, Belden C. 2011. *Ravished by Beauty: The Surprising Legacy of Reformed Spirituality*. Oxford: OUP.
Letham, Robert, and Donald Macleod. 1995. Is Evangelicalism Christian? *Evangelical Quarterly* 67: 1.
Luther, Martin. 1960:1538. *Works*. Ed. Jaroslav Pelikan, vol. 2. Saint Louis: Concordia.
Milton, John. 1667. *Paradise Lost*. London: Peter Parker.
More, Thomas. 1965:1516. *Utopia*. In *Works,* vol. 4. New Haven: Yale University Press.
Owen, John. 1652. Sermon 9: Christ's Kingdom and the Magistrate's Power. In *Works,* vol. 8. London.
Phillips, Doug. 2004. Is it Un-Christian to Hunt and Sell Fluffy Animal Pelts? Posted November 5, 2004, *The Vision Forum*, http://www.visionforum.com/. Accessed 3.5.2006.
Pink, Arthur W. 1981. *Gleanings in Genesis*. Chicago: Moody Press.
Plato. 1988. *The Law*, trans. T. L. Pangle, book VII. 824a. University Chicago Press.
Plutarch. 1909. De Sollertia Animalium. In *Works*, vol. 2, 959 c-e. NY: Kelmscott Society.
Rigney, Joe. 2025. *The Sin of Empathy*. Canon Press.
Salt, Henry. 1914. *The Blooding of Children*. https://www.henrysalt.co.uk/library/essay/the-blooding-of-children/. Accessed 27 Feb 2025
Sampson, Philip. 2001. *Six Modern Myths*. Leicester: IVP.
Sassoon, Siegfried. 1928. The Old Huntsman. In *The Old Huntsman and other poems*. New York: E P Dutton.
Schochet, Elijah Judah. 1984. *Animal Life in Jewish Tradition*. New York: KTAV.
Scully, Matthew. 2002. *Dominion*. New York: St Martin's Press.
Seidenberg, David Mevorach. 2019. Veganism and Covenantalism. In *Jewish Veganism and Vegetarianism,* ed. Jacob Ari Labendz and Shmuly Yanklowitz, 161–194. New York: Suny Press.
Sewell, Mary. 1858. *Homely Ballads for the Working Man's Fireside*. London, Smith & Elder.
Shakespeare, William. 1623a. *Titus Andronicus*. London: Edward Blount.
Shakespeare, William. 1623b. *The Rape of Lucretia*. London: Edward Blount.
Shakespeare, William. 1623c. *As You Like It*. London: Edward Blount.
Sidney, Philip. 1999:1580. *The Old Arcadia*. Oxford: OUP.
Spurgeon, C. H. 1915. Sermon 3472: A Solemn Deprival. In *Metropolitan Tabernacle Pulpit,* vol. 61. London: Passmore & Alabaster.
Spurgeon, C. H. 2003. *Morning and Evening*. Crossway: Wheaton.

Stevens, Jenny. 2012. Wildlife Official Shoots Lion. *Independent*, 2.3.2012. http://www.independent.co.uk/news/world/americas/wildlife-official-shoots-lion-7488639.html. Accessed 6 May 2017.

Stoddard, Ed. 2008. *Passion for Guns Has Roots in Hunting*, 13.3.2008. http://mobile.reuters.com/article/idUSN0725100420080313. Accessed 30.4.2016.

Stubbes, Phillip. 1877:1583. *Anatomy of the Abuses in England in Shakspere's Youth*. London: Furnival, Turner & Co.

Thomas Aquinas. 1924. *Summa Contra Gentiles*. London: Burns and Oates.

Tillotson, John. 1694. *Six sermons*, Sermon 1. London: B. Aylmer and W. Rogers.

Vantassel, Stephen. 2009. *Dominion Over Wildlife* Eugene: Resource Publications.

Vining, Wayne. 2007. Christian Bowhunters of America http://www.abpnews.com/691.article, 26.6.07. Accessed 4 Aug 2008.

Woolf, Virginia. 1943. *Three Guineas*. London: Hogarth Press.

CHAPTER 8

Animals in Warfare

The previous chapter explored the transformation of "conducting the choir" into "hunting the singers". This chapter takes up that theme by tracing the connections between hunting and warfare, leading to an exploration of the use of animals in warfare more generally.

Pre-twentieth-century war was impossible without animals, and hunting provided the training necessary for their military use, as well as accustoming the prospective soldier to blood and suffering. In particular, horses were essential for warfare. Nonconformists therefore found it notable that biblical texts forbade the king of Israel the ownership of many horses, denying to ancient Israel the most effective weapon of their age.

Today, most animals exploited in warfare are for the *development* of weaponry, rather than *as* weaponry. There is no technique of burning, poisoning, or mangling bodies which has not first been tried on animal flesh. Without animal experimentation, modern weapon development would be seriously hampered. Most discussion of animal experimentation understandably focusses on their medical use and on consumer testing. Less widely researched is their experimental use for military purposes. Unlike some medical and commercial experimentation upon animals, their military use is usually poorly regulated; and the language of warfare

P. J. Sampson, *Nonconformist Perspectives on Animals and Language*, The Palgrave Macmillan Animal Ethics Series,
https://doi.org/10.1007/978-3-032-12158-5_8

is used to legitimate cruelties towards them which peacetime normally prosecutes as unlawful.

This chapter explores the nonconformist discussion of animals in the context of warfare, cruelty and human oppression. As it largely predates the institution of the laboratory, this literature foregrounds ethical issues in everyday contexts which might otherwise appear as mere backdrops to the laboratory stage. Our attention is drawn to broader range of questions and lines of enquiry. These include consequences for both the contemporary exploitation of animals in warfare, and for the conduct of war more generally, including the "military-animal industrial complex " which is founded in the exploitation of animals (Nocella et al. 2015).

Making War

As we saw in the previous chapter, before the twentieth century, animals were integral to the reproduction of warrior masculinity through blood sports, including hunting. In early nineteenth-century Parliamentary debates about banning animal baiting and fighting, the minister for War and several MPs with a military background, spoke against reform on precisely the ground that it would weaken the fighting spirit of British soldiers. For the wealthy, it was hunting that reproduced warrior masculinity. The human relationships with horses, prey, and other animals such as hunting dogs were all oriented around developing warrior skills, hardening the heart against pity for the enemy, and accustoming men to the blood and gore of the battlefield. Modern military training no longer uses blood sports, but has found other techniques of cruelty to produced "efficient killers who will feel less compunction about attacking other humans" (Sorenson 2015, 34). Live animal training still familiarises men with blood and gore, fitting them for a "hyper-masculine culture" which may be "openly hostile to expressions of compassion" (Goodman 2015, 50–1).

Just as nonconformist discourse resisted all the aspects of warrior training arising from blood sports, so it is also incompatible with the animal cruelty in modern military training. I will look first at its resistance to the use of animals in producing the instruments of war; and then at the reproduction of the warrior spirit.

Multiplying Horses

The biblical book *Exodus* records that, some 3,500 years ago, the people of ancient Israel left Egypt. They were initially led by Moses, a charismatic prophet and lawgiver. As a theocratic society, they had no king, but Moses anticipated that in due course they would appoint one. He set out the non-negotiable requirements when they did so. These included that: "he shall not multiply horses to himself, ... Neither shall he multiply wives to himself, that his heart turn not away: neither shall he greatly multiply to himself silver and gold" (Deut. 17.16–17; see also 1 Sam. 8).

Now, everyone can think of reasons why polygyny and conspicuous wealth might be prohibited. But why should the king not multiply horses? Modern commentators make little of this. Of the more than 150,000 sermons on Sermoncentral.com, two comment on the prohibition on horses, neither substantively. However, before the twentieth century, most nonconformists were aware of the long history of the use of animals in preparation for war, and they knew the significance of horses.

In the Ancient Near East, horses meant only one thing: warfare. Indeed, they were the height of military sophistication, "an engine of war" (Borowski 1998, 100, 106). In particular, they can pull chariots to create a deadly, fast-moving military weapon, and had enabled the Hyksos and the Egyptians to conquer surrounding nations in the second millennium BC.

Dwight Eisenhower (1961) doubted that the military-industrial complex , the "conjunction of an immense military establishment and a large arms industry", predated the twentieth century. Yet the connection between military and manufacture was already established three millennia ago, with substantial investment in making and maintaining weapons, not least chariots; horses were also an essential component in a *military-animal* manufacturing complex. The Israelite leader Samuel prophesied that, in reality, their king would ignore the prohibition on horses and "will take your sons, and appoint them for himself, for his chariots, and to be his horsemen; and some shall run before his chariots. And he will ... set them to ear his ground, and to reap his harvest, and to make his instruments of war, and instruments of his chariots" (1 Samuel 8.10–12). And so it proved. King Solomon is recorded as having "forty thousand stalls of horses for his chariots, and twelve thousand horsemen"; taken at face value, a huge cultural investment in warfare for a small Ancient Near-eastern nation (1Kings 4.26). Horses remained important in warfare until relatively recently: eight million were killed during the First World War alone. Michael Morpurgo's *War Horse* movingly narrates the experiences of Joey, a horse who served in the army. By that time manufacture had been industrialised, and the horses pulled field guns, not chariots; war-horses contributed to the successful deployment of artillery, driving in turn a demand for more armaments from the now military animal-industrial complex. New relationships with animals were forged. Military leaders benefited in fire-power, and industry benefitted in sales and innovation; horses suffered and died alongside soldiers on the battlefield, as they had always done.

Israel's kings were forbidden ownership of many horses; had they obeyed the prohibition, ancient Israel would have been denied the use of the best weapon of their age, and hence the means of conducting aggressive warfare. However, as the case of Solomon shows, being forbidden ownership of horses was not the same as not owning them. What, wondered bishop Thomas Sherlock in the mid-eighteenth century, could have tempted Moses to "forbid the princes of his country the use of horses and chariots for their defence? Should such a law be proposed for France or Germany at this day, what would the world think of it?" (Coke 1801, Deut. 17.16). Sherlock assumes that the reader will recognise this as a rhetorical question. Everyone knew that the "world" of the military-animal manufacturing complex of the day would think very poorly indeed of such a proposal. But the nonconformists thought otherwise. They read that the prophet Hosea foresaw a new covenant with both humans and animals which banished weapons and warfare, bringing peace to both: God will make "a covenant for them [humans] with the beasts of the field, and with the fowls of heaven, and with the creeping things of the ground: and I will break the bow and the sword and the battle out of the earth, and will make them to lie down safely" (Hos. 2.18). For animals to "lie down safely", humans must stop waging war, then, as now. Micah (4.3) anticipated this peace of God when strong nations "shall beat their swords into plowshares, and their spears into pruning-hooks: nation shall not lift up a sword against nation, neither shall they learn war any more". Micah associates: learning peace not war, abandoning the weapons of war, and using labour to change those weapons into agricultural instruments. It is worth noting that weapons are not converted into pole axes or slaughter knives, but ploughs and pruners. The end of the suffering of animals in the military-animal industrial complex evidently coincides with the end of their suffering in the industrial-meat complex. A subject which will be taken up in the next chapter.

The nonconformist commentary on the ban on horses sometimes took on a distinctly pacifist tone. Indeed, the pacifist movement of the late nineteenth century was dominated by nonconformists. But times have changed. Horses, hunting, and blood sports have long been replaced by more scientific cruelties; and the evangelicalism which has replaced nonconformist discourse is not known for its pacifism. So, we pass through the military battlefield and the civilian industrial context of yesterday to the laboratory door of today.

The Language of War

The discourse of warfare and the dominant discourse of animals cannot be separated. In both, animals suffer symbolic violence in their representation; they also suffer manifest, subjective violence to their bodies and minds. Each reinforces the other (Žižek 2008, 11).

Animals suffer symbolic violence in war narratives which draw upon animal tropes brutalising and weaponising their lives and deaths. Phrases such as "fighting tooth and nail", "savagery", "war-hawk", "dogs of war", "a meat-grinder battle", and the "fresh meat" of novice troops, all distance humans from the in-humanity of human war by projecting our violence upon animals. Adversaries become "human animals", and war a "cock-fight", weaponising what it is to be a non-human animal; human suffering is displaced onto animal suffering, enabling warrior masculinity to deny its vulnerability. The emblematic use of the wolf, lion, eagle, and horse embodying fierceness, callous indifference to suffering, and the cunning of warfare, locates violent human qualities within animals which they may not actually possess. Such language exculpates humans from the in-humanity of warfare. Even in peacetime, the language of war legitimates violence against animals which would normally be called cruelty. Thus we speak of the "war against disease" to justify medical experiments on animals; the cosmetic industry's "war against wrinkles" entails painful safety testing (Reynolds 2023).

Yet animals don't go to war against one-another. As Augustine (1972, 12.23) observed, animals "live in greater security and peace among their own kind" than men, and "not even lions or serpents have ever carried on among themselves the kind of warfare in which men engage". Nonconformist discourse portrays animals as innocent, fitted to the manifestation and praise of God, and fierce only consequentially upon human sin. The very opposite of their appearance in war narratives.

Moreover, nonconformist discourse speaks of human vulnerability which is entirely suppressed in warrior masculinity, including a vulnerability to the very animals we have harmed. Animals and humans alike suffer both symbolic and manifest violence; all are vulnerable to war, as to plague and famine (Jer. 34.17). All can be shot, blown apart, dismembered. The psychological effects of war on animals is only recently being recognised (Salter 2015, 11). When human vulnerability is hidden behind a facade of warrior masculinity, it is harder to have compassion for the suffering of the animals we use as weapons against other humans (Stone 2018, 146).

Animals as Weapons

Two of the best known portrayals of the horrors of modern war are Goya's *May Second 1808* and Picasso's *Guernica*.[1] Both portray injured horses gazing at us in anguish, implicating the viewer in the human carnage around them. Animal mutilation is iconic of human cruelty, and comes to represent it; war has an ability to reduce even the noblest creature to suffering flesh. Horses may be emblematic of the use of animals as weapons, but humans have exploited many other species in war, especially dogs and birds.

Among the earliest examples of the weaponising of animals is that of Samson tying burning firebrands to the tails of foxes to destroy the cornfields and olive groves of his enemy (Judges 15.4–5). Dogs have long been bred for their fierceness. From Attila the Hun to Napoleon, from the Third Reich to the US troops in Abu Ghraib prison, war dogs have savaged or terrified the enemy.

Traditionally pigeons have been used to carry messages in warfare. But birds have other uses. In 1267, the sheriff of Essex was accused of plotting to release flying cockerels carrying incendiary bombs over London (Jones 2012, 295). Of course, modern industry can do better than the sheriff of Essex. The use of animals as living incendiary bombs was systematically developed by the U.S. military during World War II. Like the sheriff, they favoured flying bombs, but in their case the plan was to release flocks of

[1] Images are readily available on-line.

up to 5000 bats, each surgically fitted with a small incendiary device on a timer; the bats would shelter under the eaves in an enemy town, creating thousands of simultaneous house-fires (Glines 1990; Salter 2015, 10f). Modern psychology can also assist. During the second world war, the behaviourist B. F. Skinner (1960, 28–9) pioneered "Project Pigeon". He proposed that trained birds would guide missiles to their target. Anticipating that some people might be a little squeamish at this cruelty, he observed that the "ethical question of our right to convert a lower creature into an unwitting hero is a peace-time luxury". The language of warfare authorises suffering we might otherwise consider unethical.

Inside the Laboratories

Now, bats implanted with incendiary bombs and pigeons able to guide missiles are not found in nature. B. F. Skinner and the U.S. military weaponised their animals in laboratories, using the industrial technologies of modern science. War-horses and the dogs of war are also human products, unknown in nature. Long before the existence of laboratories, people experimented to breed and train them for war. Blood-lines were selected for fierceness, strength, and endurance. Even a carrier pigeon is a human creation, the product of many years of experimentation and selection. Flying cockerel bombs as envisaged by the sheriff of Essex, would need robust birds, carefully selected to carry incendiary material without tiring. We are nowadays accustomed to think of experimenting on animals within the walls of the science laboratory. But the development of fierce traits in horses or dogs were no less experimental because they took place on farms rather than in the lab. No less manufactured by being made in barns rather than the factories of industry. Before the nineteenth century, such research and manufacture was part of animal husbandry, and relied upon the expertise of farmers rather than the scientist. But whether in the thirteenth century or the twentieth, it was experiments which transformed animals into weapons, and it was systematic, studied production which made them available for warfare. As I have argued in previous chapters, nonconformist discourse resisted breeding animals for fierceness, and protested at cruel animal contests. This already ran contrary to the weaponisation of animals, as military protest against legislation to ban bloodsports attests. Animals, in nonconformists discourse, were for the disclosure and praise of God, and should not be bred for the promotion of violence.

Beyond the Lab Door

Early experimental methods were crude, and real "progress" had to await the scientific advances of the twentieth century. Modern industry has made possible the weaponisation of animals on a scale unimagined by previous generations. Many different species can be exploited, from dogs to dolphins, with a greatly expanded potential for cruelty. Moreover, animals know no respite, even in peacetime.

Contemporary studies of the use of animals in war have investigated the nature and extent of scientific experimentation; but most nonconformists wrote before the nineteenth century, when there *were* no laboratories. Animals were bred and tested for fighting prowess in a variety of settings, from village competition to contests in the Royal Court. Nonconformist discourse draws our attention to this wider context for animals in war.

In fact, even modern weapons-testing crosses the boundary of the lab door. Todays armaments are also tested in the field and show-cased by actual military engagements overseas, advertising national defence industries. In thus perfecting their destructive capacity, vast numbers of domestic and wild animals are blown apart, burned to death, or die of starvation following habitat loss. Aside from this collateral damage, abandoned farm, zoo and domestic animals die of neglect and starvation; yet others are used as target practice by bored soldiers (Andrzejewski 2015, 91f). This is not new. In 1613, George Abbot (1613, 453) asked rhetorically of warfare, "are not the horses wounded, and perhaps slaine in the fight? Or other cattell burnt in the stall?". Abbot was commenting on *Jonah*, a text in which the ancient city of Ninevah was spared destruction, partly on account of God's pity for the cattle living there. In the eighteenth century, John Wesley was well aware that animals will die when a city is overthrown, and also noted that Nineveh was spared on account of the innocent animals. Wesley (2010:1765, 765) has God rebuke Jonah, saying of the animals in the city, "my tender mercies are and shall be over all my works. If thou wouldest be their butcher, yet I will be their God". For nonconformists, animals are to manifest and praise 'their God in song, not to be butchered in warfare or to participate in human slaughter.

This broader context for the interaction between animals and weapons development suggests a number of issues which are currently poorly addressed, and rarely publicised: the destruction of animals during weapons-testing in the field; and the impact of capturing animals for experimental use upon both ecology and local communities. These lines of research are invisible if we conceive animal experimentation ending at the lab door.

Nonconformists addressed such issues by using categories which applied to human-animal relationships broadly, rather than to scientific protocols. They asked not about the experimental use of animals in the artificial context of the laboratory, for there were none. Rather they asked questions about the instrumental use of animals in everyday life. Questions about the purpose of the animal creation, the boundaries of human dominion, and the lawfulness of using animals to train warriors and to prepare for war. I will look at their responses to such issues, firstly in the context of the weaponisation of animals, and secondly in the context of their use to train warriors.

Nonconformist Discourse of Warfare

John Calvin (1860:1564, 11.6) commented on the ban upon horses in ancient Israel, saying that it was to prevent the "more studied modes of warfare which were in use among profane nations". Calvin knew that horses must be studied, and trials made of their utility before they can be exploited in battle. Cruelty is endemic to warfare, and experiments to achieve "more studied modes of warfare" are inherently liable to be cruel. For Calvin, this was not lawful. Three principles guided such judgements about human-animal relations: the purpose of animal creation; God's abhorrence of cruelty; and the constraints on the instrumental use of animals made necessary by sin. I will consider each in turn.

Purpose of Creation

For nonconformists, animals were created "to set forth the glory and magnificence of the great God" (Trapp 1868:1657, vol. 2, 484). The meaning of their existence is not anthropocentric, found in human use, let alone in being weaponised. Moreover, this not only applies to the animals we might find congenial. Spurgeon (1886, 148.10) discovered the praises of God in "the buzz of "the blue fly which singeth in the window pane'". We may like bats even less than blue flies, but their having been created to praise God should make us hesitate before using them as flying incendiary bombs. Nonconformist discourse provided few resources to authorise B. F. Skinner's assertion that the ethics of weaponising animals is a "peace-time luxury"; it did provide many ways to say that animals should not be weaponised.

Abhorrence of Cruelty

Animals have always suffered and died in wars. Horses are big and do not die easily; disemboweled or with crushed and broken legs, they lie on the battlefield among the wounded and dying. In Homer's Iliad we read of the noble horse Pedasus which, speared in the right shoulder, "screamed aloud as it lay, groaning in the dust until the life went out of it" (Homer 1999, XVI.196). The dominant discourse of animals may regret such suffering, but it authorises it, as horses are property to serve our purposes, even on the battlefield. But for nonconformist discourse, these cruelties are deeply problematic. Not only does breeding fierceness in a warhorse subverted the meaning of his existence, it also leads to cruelty. The screams of a horse on the battlefield, "groaning in the dust until the life went out of it", do not praise God; and God abhors cruelty.

This discourse prohibiting experimentally weaponising animals is part of a broader nonconformist opposition to vivisection dating from the mid-seventeenth century. As we saw in Chapter 5, Lord Shaftesbury protested against vivisection on the grounds that the animals are "His creatures as much as we were His creatures" (Hansard 1879, 430). Shaftesbury was referring to experiments designed to increase our knowledge of physiology in order to benefit humans and other animals alike. He would scarcely have been less critical of experiments aimed at weaponising animals for violent mutilation and death.

Our Sins in Adam

Animals, then, were created to praise God, not to enhance human cruelty and violence. Moreover, their instrumental use by humans is constrained by the consequences of our sins in Adam. There are four salient aspects which have been discussed in previous chapters.

a. Humans sin, animals do not. Consequently, animals are always already innocent.
b. Animal fierceness and suffering are products of human sin. As Calvin (1850:1559, 65.25) observed, when animals hurt humans or fight

against each other, "this ought to be imputed to [human]... sin, because his (Adam's) disobedience overthrew the order of things".

c. It is bad enough that human sin has overthrown the order of things, making animal suffering unavoidable. To exploit such suffering is egregious wickedness.
d. Moreover, cruelty is treachery against our dominion duty. Abandoning shepherding dominion, we become tyrants.

The military use of animals offended against this understanding of sin, for it made innocent creatures suffer; exploited fierceness, the result of human sin; and was a betrayal of dominion duty. As Andrew Willet (1632, vol.1, 14) argues, "..the slaughter, and killing of beasts, upon what occasion soever... belongeth unto that bondage of corruption, which by sin was brought into the world". To breed animals to enhance fierceness exacerbates the "bondage of corruption", exploits the violent consequences of sin, and is egregiously wicked.

Deuteronomy Chapter 17 prevented the use of animals in "studied modes of warfare" by the simple expedient of banning the human possession of such animals. This same absolutist approach was adopted by Shaftesbury in the nineteenth-century vivisection debate, and he only supported reform rather than prohibition against his better judgement. Naturally, such a ban on the human possession of animals would have radical consequences for the twenty-first-century laboratory use of animals, let alone those in military experimentation where it cannot even be argued that animal suffering promotes human welfare. However, the rapid growth in alternative technologies makes such an absolutist stance more credible today than at any other time in the human history of animal exploitation.

I have so far focussed on the experimental weaponisation of animals, and the consequences flowing from this. But once they had such weapons, how did warriors perfect their weapons and train for battle?

Animals and Weapon Development

Human cultures have proven very violent towards other humans, but especially towards other animals, even using animal bodies to perfect the instruments of violence we use against ourselves. There is no cruelty practised against the enemy in war that has not first been tested to perfection on animal flesh and nerve: from bullet wound to gas poisoning, from blast injury to radiation burn, from bacterial infection to napalm, animals are the first to experience the rending and tearing of flesh employed in warfare (Linzey and Linzey 2015, S2.13–14. Andrzejewski 2015, 80ff).

It is hard to know the number of animals involved, as many governments shroud military research in secrecy, promoting a strategic ignorance benefitting those who would prefer not to know. Global estimates of those used annually in experiments run into the billions; this does not include those killed or maimed in field tests or conflicts (Salter 2015, 6f). Studies suggest that this is a substantial proportion of all animals used experimentally, whether in commercial or university settings, and that the number is rapidly growing.

Moreover, unlike other animal experimentation, researchers are often free of the independent supervision or inspection which can govern civilian standards. Even in countries with relatively high regulative standards, military settings do not offer the same level of care and pain relief as would be mandatory elsewhere. Less ethical experiments are more readily authorised, or outsourced to less regulated allies (Sorenson 2015, 33. Hamilton and Eliot 2015, 123).

Experimental testing of weapons on animals differs in one important respect from medical and consumer experiments. Most are intended to maximise mutilation and incapacitating agony, not ameliorate them. Medical protocols appear strict by comparison, as the utilitarian calculus of suffering they employ is much harder to apply to military uses which aim to increase mutilation and death rather than alleviate it. But weapons development is not a modern invention. Before the twentieth century, warriors also needed to train, and to develop their weaponry before going to war. Of course, they had to experiment in the field, rather than the laboratory. In fact, they experimented literally in the field; as we have seen, hunting was the principle means for practicing horsemanship, developing strategy, refining weaponry, and preparing for bloody horror.

Hunting as training for warfare has been common since antiquity. Of course, you need leisure and wealth to hunt with horses, but even the poor could practice on rabbits and "vermin". In 1528, Baldesar Castiglione (1967:1528, 63) affirmed that many sports "are closely related to arms... hunting is the most important, since in many ways it resembles warfare". In Tommaso Campanella's utopia of 1602, boys prepared for war by exercises in military tactics and in hunting (1886, 1602, 240–1). Indeed, as recently as 1910, Theodore Roosevelt observed that nothing is better than hunting "as a training school for war" (Herman 2007, 66).

Hunting used the bodies of animals to refine the design of knife, bow, arrow, and spear. It also trained the warrior in the strategies of ambush and pursuit, as well as the tactical use of weapons in disabling, killing, and dismemberment. Using a bow or spear while riding at the gallop is a difficult art requiring regular practice. Should it be a penetrating or flesh cutting arrow? How hard to thrust without the spear twisting in the grip against bone? These were vital questions if a warrior was to survive on the battlefield, and the skills were developed on the living flesh of hunted animals. Indeed, the very dogs and horses of war were bred from their hunting cousins. Hunting was to the pre-modern world what the military-animal industrial complex is to the twenty-first century. With few horses and no hunting tradition, faithful Israel was the exception to this rule.

Ancient Israel's ban on horses contributed to its lack of a hunting tradition; in particular, there are no *righteous* hunters in the Bible. This greatly diminished the status of hunting, and the opportunities for elites to refine weapons, or develop military skills. Thomas Adams (1630, 106–7, 116) considered it legitimate to hunt "Wilde and noxious creatures" and seems to have temperamentally favoured it as a recreation with it's "profitable side", acknowledging health benefits through exercise, preparation for war, and refreshment of the wearied mind. Yet, he says, Scripture condemns it, so "if it were so hatefull to hunt Beasts, what is it to Hunt men?". He had in mind the oppression of the poor, but this state of mind is not conducive to war.

Training Warriors

Warriors not only need effective weapons and skill, they also need to become accustomed to the bloody sights of the battlefield. Blood sports encouraged indifference to bloody suffering among the labouring classes. Cockfights were said to inspire the spectator with "martial virtues, so impressive was the aggression, fury, and courage of the animals" (Perkins 1999, 8). Hunting did the same for elites, with the added benefit of overcoming any reluctance to thrust steel into living flesh. As Plutarch (1909, 959 c-e) observes, hunting is a suitable training for combat because it inures men to the sight of "wounds and blood of slaughtered beasts", at length leading them to hard-heartedness, pitilessness, and to rejoice in slaughter. Porphyry (2000, 3.20.7) argues that hunting makes the warrior "impassive to pity", a valuable quality on the battlefield.

Nowadays, military training takes a more scientific approach to training personnel to the pitilessness needed in war. But the use of animals remains. Like humans, animals are made of flesh and bones that can be mutilated, and can similarly suffer. For example, human effigies can be stuffed with animal intestines to give a more realistic feel for bayonet practice; and shooting live pigs accustoms modern military personnel to the agony and trauma of war wounds (Goodman 2015).

Nonconformist discourse speaks of kindness as a Christians virtue; hard-heartedness was not a quality greatly valued. The nonconformist role model was not the hunter, but the shepherd. But kindness and tenderness of heart are not much prized in Army Basic Training. This implicit incompatibility between nonconformist discourse and warfare matured in the nineteenth century with the foundation of the Peace Society in 1816, as an evangelical response to the Napoleonic wars. The nonconformist peace movement remained prominent until the turn of the twentieth century (Brown 2003, ch.3).

When William Wilberforce supported a bill to outlaw blood sports, militarily minded Parliamentarians argued against a ban; such soft-heartedness threatened the "martial ardour of the nation" and British success in war (Parliamentary Register 1802, 295f, 311). Modern training of military personnel no longer requires the hunt or the cock-pit, and the experimental use of animals occurs in the clinical setting of a laboratory. But if inflicting suffering on animals hardens the heart and breeds a cruel disposition, what are the consequences beyond the artificial boundary of the laboratory door? Does the same happen to animal experimenters? What is the impact on their families and their immediate community? It is known that intra-familial and community violence increases when, usually male, employees routinely exercise violence to animals (Fitzgerald et al. 2009, Jacques 2015, Linzey 2009). There is no reason to believe that the violence against animals found by undercover investigations at a range of civilian laboratories will not also occur in military facilities; indeed, the secrecy and lack of regulation make this more likely rather than less. This suggests an important research question concerning the spread of violence to *civilian* settings as a result of the experimental use of animals aimed at perfecting military violence.

The Peaceable Kingdom

The modern Talmudic scholar Emmanuel Levinas (1969, 21) observed without approval that the state of war suspends not only peace, but also morality. As we saw, B. F. Skinner agreed in the case of the weaponising of animals, although he did not share Levinas' reservations. In this chapter, we have seen the nonconformist insistence that morality must prevail, even if this involves depriving the warrior of both his cutting-edge weaponry and the training needed to use it. This might involve sacrifice, and speaks to the contemporary debate about animals and warfare. In particular, it draws attention to a number of questions which are currently poorly addressed, and rarely publicised.

Though the experimental use of animals in laboratories is appalling enough, and warrants the closest scrutiny possible, the death and mutilation of animals also occurs beyond the laboratory door. How many animals are killed or mutilated in the field, both in training and testing? How many die as "collateral" damage during warfare itself, both "wild" and domestic? If the scale and nature of the casualties were known, it would at least force explicit reflection upon how much suffering we are prepared to tolerate.

Cultural institutions have a role here. The numbers and nature of casualties are rarely reported by war correspondents. With few exceptions, animal casualties are almost entirely absent from action or war films. Currently, the general public is largely unaware of this suffering, apart from reports of zoo animals eaten or starving; or accounts of courageous individuals who rescue abandoned companion and domestic animals. If a fraction of the ingenuity, invention, and resources dedicated to the weaponisation of animals were committed to resistance and reporting, what would be the response of the general public? What safeguards could be put in place to account for these animals and ease their plight? Animal suffering in war will only end when war does. But it can be reduced.

This and the previous chapter have looked at two of the areas most associated with animal cruelty and extreme suffering: hunting and warfare. The next two chapters investigate the third such area: eating animal flesh. Those Chapters look first at the nonconformist discourse of meat-eating. Chapter 9 will trace its transformation into *carno-theology*, an innovative discourse developed by the new anthropocentric evangelicalism to authorise cruelties which might otherwise appear reprehensible.

References

Abbot, George. 1613. *An Exposition upon the Prophet Jonah*. London: Richard Field.

Adams, Thomas. 1630. *The Workes*. Thomas Harper for John Grismand.

Andrzejewski, Julie. 2015. War: Animals in the Aftermath. *In Nocella* 2015: 73–100.

Augustine of Hippo. 1972. *City of God*, Trans. Henry Bettenson. Penguin.

Borowski, Oded. 1998. *Every Living Thing: Daily use of Animals in Ancient Israel*. London: Altamira.

Brown, Heloise. 2003. *The Truest Form of Patriotism: Pacifist Feminism in Britain, 1870–1902*. Manchester University Press.

Calvin, John. 1850:1559. *Commentary on Isaiah*. Edinburgh: CTS.

Calvin, John. 1860:1564. *Commentary on Joshua*. Edinburgh: CTS.

Campanella, Tommaso. 1886:1602. The City of the Sun. In *Ideal Commonwealths*, ed. H. Morley. London: Routledge.

Castiglione, Baldesar. 1967:1528. *Book of the Courtier*. Harmondsworth: Penguin.

Coke, Thomas. 1801. *Commentary on the Holy Bible*. London: Whitfield, vol.1.

Eisenhower, Dwight D. 1961 *Military-Industrial Complex Speech*. https://avalon.law.yale.edu/20th_century/eisenhower001.asp

Fitzgerald, A. J., L. Kalof, and T. Dietz. 2009. Slaughterhouses and Increased Crime Rates. *Organization & Environment* 22: 158–184.

Glines, C. V. 1990. The Bat Bombers. *Air Force Magazine* 73: 10.

Goodman, Justin R., et al. 2015. Front Towards Animals. *In Nocella* 2015: 37–54.

Hamilton, Bill, and Elliot M. Katz. 2015. The Future of War and Animals. *In Nocella* 2015: 117–126.

Hansard. 1879. *House of Lords Debate*. 15 July 1879, vol. 248, 419–36.

Herman, Daniel Justin. 2007. From Farmers to Hunters. In *A Cultural History of Animals in the Age of Empire*, ed. Kathleen Kete, 47–71. Oxford: Berg.

Homer. 1999. *The Iliad*, trans. Samuel Butler. London: Dover.

Jacques, J. R. 2015. The Slaughterhouse, Social Disorganisation, and Violent Crime in Rural Communities. *Society and Animals* 23: 594–612.

Jones, Dan. 2012. *The Plantagenets*. London: HarperCollins.

Levinas, Emmanuel. 1969. *Totality and Infinity*. Pittsburgh: Duquesne University Press.

Linzey, A. 2009. *The Link between Animal Abuse and Human Violence*. Brighton: Sussex Academic Press.

Linzey, A., and C. Linzey, eds. 2015. *Normalising the Unthinkable*. Oxford: Oxford Centre for Animal Ethics.

Nocella, Anthony J., Colin Salter, and Judy K. C. Bentley, eds. 2015. *Animals and War: Confronting the Military-Animal Industrial Complex*. Lanham: Lexington Books.

Parliamentary Register. 1802. House of Commons, vol. 3, May 24 1802.

Perkins, David. 1999. Animal Rights and 'Auguries of Innocence.' *Blake/an Illustrated Quarterly* 33 (1): 4–11.

Plutarch. 1909. De Sollertia Animalium. In *Works*, vol.2. New York: Kelmscott Society.

Porphyry of Tyre. 2000. *On Abstinence from Killing Animals*, trans. Gillian Clark. London: Duckworth.

Reynolds, Lesley 2023. Serum Vs Botox: Which Wins the War Against Wrinkles? *Express* 30.12.2023. Accessed 20 Mar 2025, http://www.express.co.uk/life-style/style/451170/Serum-Vs-Botox-which-wins-the-war-against-wrinkles.

Salter, Colin. 2015. Introducing the Military-Animal Industrial Complex. In *Nocella*, 1–18.

Skinner, B. F. 1960. Pigeons in a Pelican. *American Psychologist* 15.

Sorenson, John. 2015. Animals as Vehicles of War. *In Nocella* 2015: 19–36.

Spurgeon, C. H. 1886. *The Treasury of David*. NY: Funk and Wagnalls.

Stone, Ken. 2018. *Reading the Hebrew Bible with Animal Studies*. Stanford University Press.

Trapp, John. 1868:1657. *Commentary on the Old and New Testaments*. London: Dickinson.

Wesley, John. 2010:1765. *Notes on the Whole Bible*. NY: Wesleyan Heritage Publishing, Jonah ch. 4.

Willet, Andrew. 1632. *Hexapla in Genesin*. London: Thomas Man et al.

Žižek, S. 2008. *Violence: Six Sideways Reflections*. New York: Picador.

CHAPTER 9

Eating and Singing

The previous two chapters discussed hunting and warfare, practices associated with extreme suffering. However, both are rivalled by the practices of cruelties and death which constitute the modern industrial-meat complex. This chapter traces the background to that industry in earlier patterns of subsistence farming, and the nonconformist response to changing social and technological conditions.

Animals in the Bible and in Everyday Life

In nonconformist discourse, animals are to manifest and praise God; Eden had a vegetable diet. Yet, cruelty and suffering are ubiquitous, and it is obvious that they prevent creatures from praising. This was nowhere clearer than in the animal husbandry of subsistence farming which required slaughter and meat-eating for survival. How should humans conduct the song of a groaning creation? In particular, what is the place of animals in the everyday life of subsistence farming communities? How should they be cared for; how should they be killed; and how should they be eaten. How, if we eat the choir, can we also conduct the song?

Nonconformist discourse addresses these question through a biblical narrative in which they recognised their own farming practices. This

P. J. Sampson, *Nonconformist Perspectives on Animals and Language*, The Palgrave Macmillan Animal Ethics Series,
https://doi.org/10.1007/978-3-032-12158-5_9

distinctively opened a moral sphere for the treatment of animals, reproduced every time they read about them in their bibles or heard about them in a sermon. The resultant discourse, suited to the pattern of meat-eating within which it was created, framed their treatment of animals, and subverted the dominant discourse of flesh-eating at a number of crucial points.[1] Cruelty was condemned, inconveniently for practical animal husbandry. Moreover, the pleasures of meat-eating were tainted by the sin which had inaugurated it, despite a positive assessment of culinary pleasure generally. Although nonconformist discourse always remained a minority way of speaking, it proved disproportionately influential (Sampson 2018).

By the mid-nineteenth century, patterns of flesh-eating in England were changing as a result of urbanisation and the industrialisation of meat production. The exploitation of animals was no longer necessary for the provision of food, but it became a foundational and central element of the animal–industrial complex of carnism (Salter 2015, 16, 20). Daily experience of animals became increasingly out of step with the biblical narrative, and nonconformist discourse decreasingly relevant to the new, carnist patterns of eating. Some nonconformists attempted to address this, but were ultimately unable to meet the challenges they faced. In this chapter, I trace these changes, and the silencing of a nonconformist voice during the emergence of the modern meat industry.

Dietary Context

The agricultural practices of nonconformists up to late nineteenth century were obviously not the same as those familiar to biblical authors in the ancient near east (ANE), but there were many similarities. In particular, both the biblical narratives and the nonconformist experience shared a daily familiarity with animals. Indeed, Ken Stone (2018, 4, 23, 45) has argued that neither the biblical text nor Christian religious experience would exist in anything like their current form without their animals. When William Cowper (2013 [1785], VI. 413–415) wrote in 1785 that

[1] I shall use the term "flesh-eating" to describe the activity of eating animal flesh, regardless of cultural context. Meat-eating is generally reserved for the practices associated with traditional animal husbandry, rather than modern carnism. Nonconformists often used the word "meat" for food generally, following the 1611 translation of the bible with which they were familiar; eg Gen. 1.29–30, where a vegetable diet is called "meat".

animals "live, dependent upon man,... at his crib, some beneath his roof", he was describing his experience of both contemporary life and his reading of the bible (Is. 1.3, Lk. 2.7).

This homology of experience and reading enabled nonconformists to recognise themselves in biblical texts. For both the text and its nonconformist reader, animal husbandry was practical farming before it became the source of spiritual symbols. Four major features of practical animal husbandry were held in common until the advent of its industrialisation:

a. Animals have an integrity of body oriented to their manifestation and praise of God. In practical terms, they were a valued part of the household, kept primarily for their milk, eggs, wool, hair, dung, and labour. Our calling to conduct their song framed their care. Slaughter was a part of herd management in conditions of relative scarcity, largely restricted to unproductive, aged, and male animals which could not be overwintered (Borowski 1998, 52f, 57f, 231f). Feasting on flesh was socially sanctioned only on festivals and special occasions; excess flesh-eating did not become the norm. Thus, early modern readers of the parable of the Prodigal Son understood that the eldest son's complaint against his father was unjustified; he had never been given even a small goat to "make merry" with his friends, not because his father was a miserly killjoy, but because flesh consumption was rare (Luke 15.11–32). Richard Baxter's view that the Patriarchs ate flesh only at a feast or sacrifice was compatible with the experience of most of his seventeenth-century parishioners (Baxter 1691, 4.3). As late as 1791, John Oswald (1791, 15) observed of the peasantry of England, "that most carnivorous of all countries, can seldom afford to eat flesh"; when they did, "the working men should have the meat", the women and children ate the gravy (Burnett 2004, 1–40).

b. The exceptions to this were the rich who regularly ate flesh, sometimes to excess. The flesh-eating culture of the rich generated both a transgressive desire, and a transgressive indifference to the source of the flesh, specifically to cruelty (Goodman and Sage 2014). Rather than leading the song, sin intervened, offering in exchange the excess pleasure of consuming the choir, like the wicked shepherds of Ezekiel 34. A shepherd must "not *feede upon* his flocke, but *feede* his flocke", says John Rawlinson (1612, 24). Consuming is the opposite of conducting. Conducting is to enable each distinct voice to contribute to the whole. Consuming is to devour, to homogenise, to silence. No excess pleasure will ever be enough to fill that silence. So we consume without limit, and nothing is enough to enable us to truly enjoy what we have.

c. Dietary disparities between rich and poor were commonplace in both the biblical texts and pre-industrial England. No doubt the imaginative poor produced tasty meals, but the temptation of excess feasting on meat was restricted to the rich. Nonconformists noted the strong biblical condemnation of excessive eating and its association with injustice (eg Jer. 5.26f, Ez. 16.49). Many added their own. Gluttons, says Calvin (1849 [1565], 16.49–50) "those who have abundance… arrogate to themselves more than is just". Richard Baxter (1678, 317) complains: How can one "member of Christ" be a glutton when "another is starving and pining in a Cottage, or begging at the door?" Some work hard and are scarce "able to get bread for their families, that they may bring in all to the *belly-god* Landlords, who consume the fruit of other men's labours upon their devouring flesh". A century later we find Hannah More (1840, 233) arguing that the rich should abstain from buying cheap meat for gravy so that the poor might buy it for sustenance. This association between gluttony, injustice, and flesh-eating endorsed the taint of the latter arising from its origin in sin.

d. Whether among rich or poor, flesh-eating involved unavoidable cruelty. In the case of bloodsports and labouring animals, nonconformists condemned gaining pleasure through cruelty. But there was also substantial overlap between blood sports and meat-eating. Hunted animals were often eaten; chasing to exhaustion and bullbaiting were both practised to tenderise the flesh. "But if our sports are destructive, our gluttony is more so, and in a more inhuman manner. Lobsters roasted alive, pigs whipped to death, fowls sewed up…" (Pope 1713). Working animals could also become flesh through abuse; as popular a song said: "When we have gall'd and tyr'd the Mule, And can no longer have the rule, We'le have the Spoyle at least" (Anon 1662, 12). The celebration of abusive power over the mule was no more acceptable to nonconformists than pleasure through blood sports. While flesh-eating was permissible, cruelty was not. "Gods mercy, care and providence … extendeth to the bodies and lives of [animals]", wrote Lancelot Andrewes (1650, 278), advocating that cattle be fed and rested.

I shall henceforth refer to this pattern of flesh consumption as meat-*eating*, a term which indexes pre-industrialised practices of animal husbandry.

The Nonconformist Discourse of Diet

The nonconformist discourse of meat-eating was little concerned with diet in the modern sense. Rather, it concerned the interconnected network of threads woven into the everyday life of both the biblical narrative and their own: the provision of vegetables to eat in Eden, to be renewed at the restoration of all things; the absence of a creation mandate for eating animals, making it tacitly transgressive without specific sanction to permit it; the necessity of meat-eating as a result of the scarcity caused by sin; minimising the cruelty this involved; restricting the addictive excess pleasure of flesh-eating which could lead to gluttony and injustice; and the impossibility of gaining spiritual merit though either legalistic or ascetic patterns of eating. As we shall see in the next chapter, these concerns are far removed from modern anthropocentric evangelicalism's focus on diet.

As meat-eating wasn't a major feature of everyday life, nonconformist discourse does not often address it specifically, except when linked with either gluttony or obtaining spiritual merit though dietary practices, both of which were disapproved. So far as I am aware, the nonconformists produced no systematic treatment of the subject until Humphrey Primatt's *The Duty of Mercy* in 1776. Rather, it was discussed by preachers and commentators as it appeared in the everyday life portrayed in biblical texts, and lived by their congregations. They distinctively sought to understand their experience of animals in a mixed community of labour, care, and herd management from within a biblical narrative which shared this experience. The congruence between the animal husbandry of the bible, and that of early modernity, generated and reproduced a moral sphere for both the treatment of animals (practical animal husbandry), and for reading about them in the biblical texts (their distinctive hermeneutic). By the time of Primatt's text, biblical animal husbandry was becoming unrecognisable in the industrialising diets of his parishioners. A national industry was emerging, with animals driven long distances on public roads to centralised slaughter in cities. Following Joy (2009), I shall refer to this new, industrialising pattern as *carnism*.

Nonconformist discourse addressed the practical issues of animal husbandry which would be faced by farming congregations. In particular, the duties that arise from an animal's ontological status as a creature of God. As we have seen, John Maynard (1688, 19–20) addressed abuses familiar to his seventeenth-century readers:

The "withholding of due and necessary food from the Beast, or over-burdening it beyond measure, is a dishonour offered to its Creator. And as these Beasts of service must not be abused, because they are the Creatures of God; … So for those Creatures which we use for cloathing, nourishment, or other use, we must take heed of all excessive abuse, and so of taking the name of the Lord in vain. Still remember, that this is created of God, and I must use it as a Creature of His…. shall I take Meat and Drink, the very Creatures and Workmanship of Gods own hand, and make them Instruments of Satan and sin, of Gluttony, of Drunkenness, means whereby to dishonour God, and fight against Him that made them?".

Maynard argues that our duties towards animals flow from their ontological status as "creatures of God". This frames their "service", which must not be abused, even if the experience of "excessive abuse" ("gluttony") is pleasurable. There are two intersecting discourses here: what animals are; and how we use them. I will treat each in turn.

What Animals Are

Animals, along with all creation, are to manifest and praise God; humans are to facilitate their song. But after the Fall, it became necessary to use animals in subsistence farming, for "cloathing, nourishment, or other use"; this meant unavoidable suffering for many, and death for some. This is nowhere clearer than when an animal was slaughtered and its flesh eaten. This obviously subverts the human calling to conduct the animal's voice, and brings an end to her praise.

Eden Was Vegan

In the Genesis narrative, God pronounces creation "good" on six occasions, "very good" after the grant of a vegetable diet to humans and all other animals (Gen. 1). George Walker (1641, 232–3) summarised the majority nonconformist view: "in the state of innocency man had no power over living creatures to kill, and eat them; neither did one beast devoure another and feed on his flesh; but the food of man was only herbes and fruits of trees; and the food of beasts and birds was the greene herb and grasse of the field".

All animals, human or not, shared a common creational space in which only plants were ontologically fit for consumption, fit to cease existing. Animals were no more fit for human consumption than we exist as food for them. Neither meat nor meat-eating existed. All creatures were members of a complex community, joined in co-dependency, singing praise and manifesting their creator according to their kind (Stone 2018, 180–1).

But we no longer live in Eden. Did animals become fit to eat in the shadow of the Fall? Did meat-eating become lawful?

Eating in the Shadow of the Fall

The theological Fall was itself represented as an act of eating a forbidden fruit; moreover, like flesh-eating, it was an act of eating that was not creationally mandated. Rather, it was transgressive, entailing both over-consumption and epicurean desire (Gen. 3.6). Thereafter, "thorns... and thistles" compromised farming, and violence grew until it "filled the earth" (Gen. 3.18 & 6.11–13). Many nonconformist considered that this violence extended to animals eating one-another, and even to anthropophagy (Sampson 2018). The Fall enabled flesh-eating to come into the world, but, without sanction to permit it, it remained ontologically transgressive. Few nonconformists thought it natural in the normative sense of *lawful*. Before the flood, said Matthew Henry (1710, Gen. 9.3): "most think man had been confined to [lawfully] feed only upon the products of the earth, fruits, herbs, and roots, and all sorts of corn and milk; so was the first grant". Adam Clarke (1831, Gen. 9.4) remained of this view in the nineteenth century: "Before the deluge... animal food was not [lawfully] in use".

Flood and Flesh

The vegetable diet of Eden changed with the Flood narrative, triggered by the violence that had filled the earth. John Trapp (1867 [1662], Gen. 9.3) called the post-diluvian world "new and much weakened"; not only was agriculture destroyed by the deluge, but "a great decay was wrought both in the earth with its fruits, and also in man's nature". This, says Henry (1710, Gen. 9.4), "rendered its fruits less pleasing and less nourishing". The survivors of the flood "should be allowed to eat flesh for the support of their lives...". The violence of slaughter and eating flesh became necessary for survival; to live, humans had to "give death" (Douglas 1999,94–6. Seidenberg 2019, 170).

But the violence of flesh-eating was transgressive, so a new "grant" was necessary to authorise it as lawful. Lacking a creation mandate such as ontologically fitted plants for consumption, this needed a social marker to indicate its sanction and its boundaries. The Noahic covenant provided this (Gen. 9.10–17). It partially recapitulates the grant of a vegetable diet in Gen. 1.29–31, but conspicuously omits the judgement that eating flesh is "good", let alone "very good". Flesh had crossed an ontological boundary; animals were permitted for human consumption; meat came into the world.

This social sanction permitting meat-eating could not make it natural in the sense of mandated at creation. Eden's "very good" plant diet had required no social marker. Meat remained un-natural, abnormal, born of sin, and *naturally* abhorrent. Meat-eating is an aberration which, according to eschatological promise, will cease at the restoration of all things, when, as Walker (1641, 232–3) puts it, "men shall not kill and eat up beasts nor beasts one another". On the new earth, there "will be no more death or mourning or crying or pain" (Rev 21.1–5). Once again, there will be no meat, only flesh; only embodied creatures who will sing songs of praise, as it was in the beginning.

The contrast between Eden's diet and flesh-eating recurs throughout the biblical texts. Blessing, praise, and celebration are often associated with a vegetable diet: examples include Eden itself; the produce of the promised land; the bloodless feast of the eucharist; and the eschatological promise of the lamb residing *with* the wolf, not *within* the wolf. Moreover, flesh-eating could scarcely be more ambivalently authorised: the notable failure to call it "good", let alone "very good" at its inauguration; its association with covenant breaking and transgressive desire; and its absence in the new earth. As a result, meat-eating carried an inherent taint of transgressivity, reflexively affirming the boundary between Eden and the sinful world (Goodman and Sage 2014, xi, xiii). In "the awful violence" of animal slaughter, says Andrew Bonar (1966 [1846], 1.15), "a testimony against sin ascends up into the ears of the Lord of Sabaoth". As Thomas Adams (1862 [1629], 137) pithily observed, "it was sin that made us butchers, and taught the master to eat the servant". Sin was integral to the genealogy of flesh-eating; sin remains the condition of its possibility.

While flesh-eating may be lawful, even necessary in conditions of scarcity, it was tainted with sin. Moreover, God's known hatred for cruelty to his creatures intensified the conflict between conducting the song of creation while exercising violence towards the singers.

Cruelty After Noah

Violence and cruelty are usually only socially sanctioned within specific circumstances: for example, punitive violence by the state, or warfare. Violence outside these circumstances is usually considered transgressive. The Noahic covenant sanctioned violence only in the specific circumstance of post-diluvian scarcity, and of herd management in subsistence farming. It did not authorise violence in excess of this. Thus Cowper (2013 [1785], VI.396, 454–8) regarded eating animal flesh as lawful since the "charter" conferred upon Noah "and in him on all mankind" but, he says, "read the instrument [Noah's covenant], and mark it well: Th' oppression of a tyrannous control Can find no warrant there... Carnivorous, through sin, Feed on the slain, but spare the living brute". He refers here to the central requirement of the Noahic covenant which concerned the way an animal was to die. In subsistence farming, stock management required that some animals have to be slaughtered to preserve the health and balance of the herd. But animals may be killed in a variety of ways. Not only does the degree of suffering depend upon the method of slaughter, but there are practical considerations. The flesh of pre-modern "food animals", including game, was often tough and indigestible. It could be tenderised by various violent practices such as running to exhaustion, beating to death with clubs, or baiting with dogs (Pinker 2012, 554). Moreover, the practical problem of preserving the flesh of a large animal once it was dead made it more efficient to carve flesh a little at a time before death. Steven Pinker (2012, 552) notes that the "cutting or cooking of live animals by traditional peoples is far from uncommon". But the social marker of the Noahic covenant prevented all such cruelties by requiring that the blood to be drained from an animal, ensuring that it is dead before it was dismembered, cooked, or eaten. "The restraint that was of eating the blood of dead beasts, declared that he would not have tyranny exercised on them while they are alive", says John Trapp (1867 [1662], Prov. 12.10). Or, as John Gill (1811 [1763], Gen. 9.4) put it, to "restrain cruelty in men", that an animal "should not be devoured alive". The very "grant" inaugurating meat-eating was contingent upon the requirement to restrain cruelty and violence—a recognition that otherwise it would be used to express forbidden, repressed desires, and to generate transgressive pleasures; in short, to reproduce the violence which had formerly filled the earth.

Cruelty and Law

The Mosaic covenant consolidated and elaborated the social sanctions of the Noahic covenant, principally to limit violence and ameliorate cruelty.[2] It permits only the consecrated killing of sacrifice, effectively preventing it from becoming routine. "The extreme sensitivity to bloodshed and loss of life, human and animal, shown throughout the book [of Leviticus] is consistent with the prohibition of profane slaughter", argues Mary Douglas (1999, 68). In particular, it declared "unclean" (inedible) those animals whose slaughter entails intense agonies, including swine, crustaceans, cetaceans, and cephalopods. Even "clean" (edible) animals may not be slaughtered or eaten if they bear the scars of abuse or mutilation. This implied very high standards of welfare, encouraged care and empathy, and outlawed the routine mutilations of carnism.

Adherence to the Law ensured that animals were cared for during life, and were not injured or mutilated before death; just as the requirement to drain the blood ensured that they were dead before being butchered, cooked or eaten. Moreover, animals may only be slaughtered at the cultic centre, which for many was distant, for some far distant, under pain of being "guilty of bloodshed" (Lev 17:3–4). To avoid injury, the journey had to be at the animal's pace, even involving carry her. This, combined with rearing the animal in the first place, encouraged close identification and empathy between man and beast, militating against casual abuse.

Some parts of some sacrificed animals could be eaten, but not the fat; moreover, her flesh must be boiled not roasted. The exception being the passover lamb which was roasted, but with intestines intact and with bitter herbs. The law permitted killing animals, but prevented the distancing and indifference which isolates modern carnism from the reality.

Animals are, said Maynard, "the Creatures of God" and their "service must not be abused". To do so, is to make them into "means whereby to dishonour God, and fight against Him that made them". The covenants authorising flesh-eating constructed a food culture which restricted cruelty and excess for most people. Hunting was disapproved, slaughter limited to the cultic centre, and meat-eating surrounded with numerous restrictions. Both the Noahic and Mosaic covenants recognise that active steps are required to encourage empathy, and prevent transgressive pleasures.

[2] I am here discussing the law as a dietary discourse, not as cultic, or religious; nor as a marker of identity.

This leads us to the discourse of using animals, and the pleasures derived from it, particularly culinary pleasure.

The Use of Animals

In the Christian history of diet, pleasure has sometimes been treated with suspicion, even as inherently transgressive, especially where ascetic deprivation is considered meritorious. But nonconformists rejected asceticism. Contrary to popular myth, neither Puritans nor other nonconformists were opposed to the experience of pleasure. In fact, nonconformist religion was characteristically experiential, or as was said in the seventeenth century, "experimental". Without experience, we cannot enjoy the benefit of religion. "Would a painted fire content you in a cold winter?" asked Anthony Burgess (1652, 126). "Would painted food satisfie your hungry appetite? No more should the meer doctrinal part of Religion, without the practical part". This applies to our experience of the manifestation of God in creation: His attributes, says Baxter (1678, 214), appear in His works and "do make them as a glass in which at this distance we must see the Creatour, and their sweetness is a drop from him, by which his goodness and love is tasted". The creatures are a "love-token from our dearest friend". We taste God's goodness in the sweetness of his living creatures and their song, not in their roasted flesh. What kind of person would treat love tokens as we do "food" animals?

Vegetarian Desire

As Burgess intimates, pleasure is an important part of eating. It began in Eden, whose vegetable diet was not simply nutritious and healthy, sustaining the body. It also revealed the goodness of God in satisfying the appetite, yielding a pleasure which overflows that of simply sating hunger. This *surplus* pleasure is always with us as humans, an epicurean enjoyment of taste, aroma, and texture which adds nothing to the nutritional quality of food (Žižek 2022, 6). As Calvin (1965 [1578], Gen 1.28) commented, Eden's vegetable diet was "abundantly sufficient for their highest gratification … a liberal abundance, which should leave nothing wanting to a sweet and pleasant life". John Yates (1622, 160) observed that the trees and plants of Eden were "as well for pleasure as for profit". Indeed, according to Thomas Adams (1862 [1629], 134) Eden was inherently for "pleasure". "Such was the Maker's bounty, not only extending to life, but to the happiness of life. Every part of earth would yield fruits enough for Adam's sustentation, this shall also for his delectation ... Our lawful pleasures are his [God's] pleasures, ... Men may eat and drink even to honest delight, so withal they worship the giver". John Wesley (1872 [1771], 2.1.5) told his congregation that food in Eden was not only "plenteous" but included "every thing that could give them pleasure". Plants, wrote John Ray (1995 [1686], vol. 3, 1179) "supply abundantly from their very wealthy store" both "basic foodstuffs" for nutrition and "delicacies" for pleasure. I labour this point only because nonconformist, especially Puritan, Protestantism has often been presented as dourly forbidding of pleasure.

Nutritional provision and the surplus pleasure of eating were bound together in the vegetable diet. Moreover, in Eden, the desire for surplus pleasure was not transgressive. As Matthew Henry notes, before the Fall, "humans had no more desire to eat flesh than a sheep has to suck blood like a wolf". The "delightful fruits of the earth" would have been preferable to flesh, says Luther (1960 [1538], vol. 1, 72; vol. 2, 134). "For Adam it would have been an abomination to kill a little bird for food". In prelapsarian Eden, abundance to sate hunger went hand in hand with the surplus desire for a good meal and its attendant pleasures. God intends us to enjoy food, not simply to be sustained by it. Neither the vegetable diet of Eden, nor its surplus pleasures and desires, required a cultural sanction to warrant them; they were creational mandates, they were "natural".

Most nonconformists considered that the continuing surplus pleasures of a vegetable diet is a trace of Eden, and is not transgressive. God provides a vegetable diet for "beasts and men", argued David Dickson (1965 [1655], 229): the "Lord's allowance for man is very large, not only for necessity, but also for delectation". Dickson regarded its "delectation", its surplus pleasure, as a gift and blessing. In this, he followed the biblical perception of Arcadian vegetable diets as unambiguous blessings, even in a sinful world (eg Num 13.23f; Deut. 8.7–9.,1 Kings 4:25, Ez 34.27). For William Cowper (2013 [1785], III.307f, 298–99), this trace of Eden can be nurtured; he envisaged restoring a rural "[p]aradise (for earth has still, some traces of her youthful beauty left)" which excludes hunters who "defile [it] with blood". For nonconformists, this "trace" is also an anticipation of the restoration of all things, when strong nations "shall beat their swords" into into agricultural instruments; not into pole axes or slaughter knives, but ploughs and pruners (Mic. 4.3). The end of the suffering of animals in the military-animal industrial complex evidently coincides with the end of their suffering in the industrial-meat complex. John Ray (1995 [1686], vol. 3, 1179) asks: "How much more innocent, neat and healthy is a table spread with them [seeds and herbs] then [sic] one with slaughtered and mangled animals?".

However, once flesh-eating became lawful, the desire for surplus pleasure, for "delectation" in eating, becomes more complex. The surplus pleasure of Eden's vegetable diet involved no suffering, but the surplus pleasure of meat-eating does, even if the Law is followed, which often it is not. So, nonconformists were faced with the question of gaining pleasure through suffering. We have already seen in Chapter 6 that nonconformist discourse condemned gaining pleasure through suffering in the case of blood sports. There, "sportsmen" (generally men) had a choice. But what of the relationship between pleasure and eating where the choice is restricted by scarcity?

Meat, Transgression and Excess

The grant to eat animal flesh originated in sin, violence, and the human renunciation of praise to God. In this there is a parallel with blood sports. The first after the Flood who crossed the ontological boundary separating praising creature from meat, who first sated his hunger on animal flesh, would inevitably have experienced meat as horror. An animal whose "gambolling" and "bounding" had formerly sung God's praise, that had given such "utt'rance as they may to ecstasy too big to be suppress'd", was now mangled and bloody (Cowper 2013 [1785], VI. 243). The Platonist philosopher Plutarch (1962, bk 3, tract 1) pictured the first who "with his mouth touched slaughter, and reached to his lips the flesh of a dead animal… that but a little before lowed, cried, moved, and saw"; this man, he says, who first "participated of the saps and juices of deadly wounds", knew fully the existential abhorrence of the act. Plutarch had in mind only the novelty of flesh-eating and the intuition into praise which all men have by common grace. When we consider that the biblical Patriarchs are presented as having a profound experience of both creation's praise, and of the Creator praised, this horror would have been multiplied uncountably. Flesh-eating constitutes the negation of praise; screaming and dead animals cannot praise. Consuming, bringing to naught, is the opposite to conducting, bringing to fruition. In nonconformist discourse, the transition from horror at tasting the "saps and juices" of flesh, to pleasure at consuming it, is deeply problematic.

The ambivalence and transgressive taint of meat-eating is clear in Richard Baxter's lament of 1691, "I am convinced that to eate flesh is lawful, and yet all my daies it hath gone, as against my nature, with some regret" (Baxter 1691, 39). We know little of the diets of nonconformists, but some such as Ray, Cowper, Wesley, and Spurgeon, if not entirely vegetarian avant le lettre, seem to have shared much of Baxter's "regret". This taint of transgressivity was distinctive to nonconformist discourse, and almost unknown among their contemporaries; indeed, it is not commonplace now.[3]

The first to eat flesh may have done so with horror, but it seems that human experience soon adjusted, and horror turned to desire. Mangled flesh became juicy, succulent steak; a new taste for flesh emerged from experiences of eating - more appealing than appalling. What should have moved us to mourn, now made us rejoice, to paraphrase Perkins (1602, 589). "For us", says Luther (1960 [1538], vol. 1, Gen 1.29), "nothing is more delicious than meat". As one twenty-first century chef put it: "When you caramelise a good piece of fish or meat, and see juices oozing out of it, it gives you satisfaction that no vegetarian cooking can match" (Hollihead 2007, 15). This transition from Eden's horror to a satisfaction occasioned by sin is illustrated by nonconformist discussion of a particular group of pericopes: those which deal with the desire to eat meat (Ex. 16, Num. 11, Ps. 78 and Ps. 106).

Craving for Flesh Pots

After ancient Israel left Egypt, they travelled through an inhospitable desert, and craved the "flesh pots" of Egypt. Some take the accounts to refer to one occasion, others to two. Whichever is the case, the desire for flesh is specific, and employs a word variously translated as "crave" or "lust" which would remind the native Hebrew speaker familiar with Torah of the *desire* for the forbidden fruit (Gen. 3.6), and of the sin of *coveting* in the Ten commandments (Deut 5.21). Calvin (1845 [1563], vol. 4, 106.15) finds in the Hebrew a wordplay which "would signify *good pleasure"* as well as *lusting*, linking the lust for flesh with the pleasure at eating it. The texts record that the craving was answered with a miraculous supply of quails. However, it seems the people's lust was insatiable,

[3] There are of course exceptions. For example, Shelley (1947, 6, 8) speaks of the "disgust [at] the horrors of the shambles", of "intolerable loathing and disgust", although this seems more aesthetic than is common in nonconformist discourse.

for they ate without restraint until the flesh came out at their "nostrils"; this excess was brought to an end not by satiety, but by a plague while the "flesh was yet between their teeth" (Num.11.20, 33). The very negative associations between craving for flesh, excess, and carnist pleasure are clear. Nonconformist discussion of these pericopes raises two separate but interconnected issues: excess; and craving or lust.

From Surplus to Excess

Whereas nutritional provision and the surplus pleasure of eating were bound together in Eden's *vegetable* diet, the surplus pleasure of *flesh*-eating is untethered from nutritional needs or health; it becomes excess. The link between meat and poor health is not a recent discovery.[4] Flesh-eating, says Luther (1960 [1538], vol 1, Gen. 1.29), results in "leprous obesity" whereas the vegetable diet of Eden had led to "physical beauty and health and a sound state of the humours". Calvin (1845 [1563], vol. 4, 106.15) more prosaically says its excess results in an "unhealthy state of the body". By this untethering, surplus pleasure becomes excess pleasure, the "excessive devouring of his owne meate", the "excessive appetite" of Israel when they lusted after flesh (Dod and Cleaver 1635, 267–8). By contrast, there are few instances of the surplus pleasure of Eden's diet becoming excessive; not many gorge on greens or pig-out on peas.

While most of their contemporaries fully accommodated to their desire for flesh, something of the taint of transgressivity persisted in the nonconformist experience; they retained a passionate awareness that the animals would never have "groaned under the killing knife and slaughter, if man had not sinned" (Walker 1641, 232–3). The "creatures suffer for our sake", as Thomas Wilson (1614, 583) put it. A man so steeped in biblical literature as Wilson can scarcely have been unaware of the trope of the suffering servant, that "Christ hath suffered for us" (eg 1 Pet. 2.21, 4.1). Meat-eating was a reminder of "what sin hath brought on the world, and what we deserve ourselves"; it was also a reminder of the sacrifice of Christ for our salvation (Baxter 1691,39). Andrew Bonar (1966 [1846], 1.15) is explicit, saying that animal sacrifice exhibits "another feature in the death of Jesus, viz., the awful violence done to one so pure, so tender, and so lovely. We shrink back from the terrible harshness of the act, whether it be

[4] It must be noted that I am referring to cultural consumption patterns, not individual eating behaviours.

plunging the knife into the neck of the innocent lamb, or wringing off the head of the tender dove". William Cowper (2013 [1785], III.108–116), unambiguously identifies the suffering, crucified Christ with the agonies of a hunted stag.

The vegetable diet of Eden evoked gratitude and praise. God had provided for all animals to enable them to manifest and praise Him. Flesh-eating, on the other hand, resonated with suffering, destruction, and death. Untethered from nutrition, flesh-eating ushers in unhealthy eating, even if not always leading to Luther's "leprous obesity". It forfeits the pleasure of praise, and in exchange gains the excess pleasure of destroying by consuming the choir. Flesh-eating savours of John Milton's devil who destroyed all pleasures other than the excess delight in destruction itself: "all pleasure to destroy save what is in destroying! Other joy to me is lost" (Milton 1667, 9.476–9).

Pleasure and Violence

Most people who eat meat but claim to abhor cruelty can experience a cognitive dissonance. Mostly the cruelty and violence is veiled in various ways (Joy and Tuider 2016, viii–xii), but veils can be fragile. There is curiously little discussion of another possibility: Milton's demonic "pleasure… in destroying", the pleasure which may be obtained by cruelty, not least by culturally sanctioned cruelty. This is historically familiar in the Roman Circus, the Inquisition, and so on. But there is a widespread assumption that modern Western cultures do not sanction cruelty in this way, and therefore eliminate the possibility of gaining pleasure by sanctioned cruelties. Meat is the obvious exception.

Socially sanctioned cruelties were well known to nonconformists. Bulls were not only baited to tenderise their flesh, the violence became a sport. Likewise hunting to exhaustion not only makes the stag's flesh more edible, but provides pleasure to the hunter. Blood sports of various kinds were part of everyday experience in the seventeenth century, and nonconformists could not avoid its reality. From the eighteenth century, animals were cruelly driven through towns in full public view. Working animals, especially in the growing cities, were abused and beaten, sometimes to death. These documented abuses are hard to understand unless the criminals enjoyed inflicting cruelty. Nonconformists protested against all these excessive pleasures in cruelty, yet could not rid themselves of the taint. Modern people no longer witness cruelty in his way, and it is relatively easy to deny its reality. Yet violence remains necessary to make meat out of living animals, and transgressive cruelties in the meat industry are well documented. Moreover, as previously, the infliction or witnessing of suffering may itself become pleasurable, even sexually charged (Eisnitz 2007; Genoways 2015;Foer 2009, 181–2, 256). The nonconformists could not avoid the reality that, when excess pleasure runs free, cruelty follows. Modern carnists can avoid it, and rarely recognise the connection between their diet and an enjoyment of cruelty which most would certainly wish to disavow.

An Anatomy of Craving

Meat-eating generates and reinforces desire for excess pleasure, an elementary imbalance between desire and satisfaction, a hungry insatiable lust which reflexively affirms the boundary between Eden's vegetable diet and the over-consumption of the Fall. Nonconformists were familiar with examples from the biblical narrative. King Solomon exhorted his son to be wise and not keep company with "riotous eaters of flesh"; and the prophet Amos criticises the rich who "eat the lambs out of the flock, and the calves out of the midst of the stall", echoing Ezekiel's wicked shepherds (Prov. 23.20, Amos 6.3–6, Ez. 34). Isaac loved "savoury" venison, and that did not go well for him (Gen. 25.28, 27.3–14). We have already seen that ancient Israel "fell a lusting" and "craved" flesh in the desert.

For nonconformists, the most familiar example of excess pleasure, from both their own experience and the biblical narrative, was that of "roast" meat. Lean meat has little taste, even less when boiled. The flavour can be improved by adding body fats, or by roasting to develop sugars of caramelisation. This is familiar in today's industrial foods which are carefully developed in laboratories to reach the "bliss point"—that perfect combination of fats and sugars which amplifies desire, and makes food irresistible irrespective of nutritional value. Indeed, the addictive quality of a desire for meat is often candidly expressed by the admission that "I couldn't give up bacon/steak/ 'bbq' etc.". Recent research points to a biochemical basis behind such addictive behaviour (Stevenson et al. 2020).

The Mosaic covenant precisely forbade the consumption of an animal's body fat, and the roasting which feed this addiction. Fat had to be burned, not eaten; and flesh was to be boiled, not roasted or caramelised (Lev 3.14–17, Lev 7.22–25, 1 Sam. 2.13–16). This ensured that the flesh was as tasteless as possible, yet retained its nutritional value. The only flesh to be roasted, that of the Passover lamb, was rendered unpalatable by "bitter herbs" and by leaving the internal organs intact (Ex 12.8–9); tradition preferred it be eaten after the appetite is satiated, not to satisfy it—a practical consideration if it did not taste good (Talmud - Seder Moed, Pesachim 70a). Moreover, the flesh of ancient animal breeds was tough and stringy, requiring long boiling to make it digestible—especially as the brutal tenderising techniques used elsewhere were forbidden. Long boiling reduces the flesh to a mush, which is why Eli's sons demanded roasted meat, not "sodden" [boiled] flesh, a "very great" sin before the Lord (1 Sam. 2.12–17, 29).

The forbidding of fat and roasting was widely read by nonconformists as restricting excess pleasure, since the fat would be the "most *attractive to the greedy*", the most "delicious", the part "many… would *much desire*" (Calvin 2018,342. Gill 1811 [1763], 7.25, citing Maimonides. Jamieson et al. Lev. 3.4–11. Wesley 1765, Lev. 3.17). While they regarded the Mosaic food laws as surpassed, nonconformists were aware of the transgressive nature of excess which had made them necessary, and the illicit pleasures it could engender. The difficulty is, says Calvin, that humans have "so perverted everything" that we are "insatiable pits" (Bouwsma 1988, 140).

A Culture of Excess

Excess and craving combine to generate pure excess pleasure which floats free of restraint and knows no limit; becomes, as Calvin observed, an "insatiable pit". The taste of flesh spawns a taste for flesh; to paraphrase Hamlet, "it grows by what it feeds on". No natural satisfaction is possible, no excess is enough. Only by the experience of eating too much can we calibrate the sufficient, by which time we have exceeded it. The "minds of men are not kept within the bounds of reason and temperance", says Calvin (1845 [1563, 78.26), "they become insatiable; and, therefore, a great abundance will not extinguish the fire of a depraved appetite". "[G]uttlers [gluttons] whom no abundance can suffice to satisfy. … although they were bursting with excess, they were not satiated; but were so inflamed by their boundless voracity" (Calvin 2018, vol. 4, Num.11.18). John Trapp (1867 [1662], Ez 34.3), addressing the wicked shepherds who eat the sheep, is characteristically blunt: "ye gorge yourselves with the best of the best"; "Ye sacrifice them; so ye pretend, but mind your own fat paunches".

Transgressive desire for more can breed a *culture* of excess. Foods are considered irresistible, not because of their nutritional qualities, but because their excess pleasure is culturally mandated as an end in itself. Even if satiation occurs, excess pleasure can continue provided the appetite is tickled, or the stomach emptied. The culturally sanctioned practice of vomiting in order to eat more was known to the ancient world, and has since been attributed to elites such as the Tudor court.[5] Vomiting was especially associated with meat-eating, and may also have served to prevent the constipation associated with meat-heavy diets. It maximises excess pleasure by subverting the natural limit of satiation. Indeed, breaking the link with digestion by vomiting produces pure excess pleasure: "They vomit so that they can eat, and they eat so that they can vomit. They don't even consider the dishes which they have assembled from across the earth worthy of digestion" (Seneca 1932, 448–9). This has widely been seen as a decadent vice, an example of Calvin's "perverted" appetite. Interestingly, Calvin recommends we turn to animals to learn moderation of excess (Bouwsma, 1988, 140).

The Birth of Carnism

The mass availability of meat and its routine consumption opened a space for transgressive excess pleasure to be normalised. Most people welcomed this as progress; by the late nineteenth century, the pleasures of carnism became the norm for an unprecedented proportion of the population in Western Europe.

The initial nonconformist response followed the same lines as their earlier view of meat-eating discussed in this chapter. They accepted its necessity, but could not stomach the abuses. However, simply preaching to a local farming congregation no longer met the challenges of urban carnism, and some tailored the earlier discursive resources to campaigning for law reform.

From the unsuccessful bill to stop bull baiting of 1800, there were a series of attempts at reform before the passage of the 1822 Act to "prevent the cruel and improper treatment of cattle"; subsequent Acts extended and refined its provisions. Those speaking in favour of reform repeatedly appealed to nonconformist discourse; two of the prime

[5] I am here referring to a sanctioned social pattern, not an individual practice.

movers, William Wilberforce and Richard Hill, were known to be evangelical Calvinists. In 1824, the *First Prospectus* of the SPCA appealed to the "great moral and Christian obligation of kindness and compassion towards the brute creation" (Linzey 2000, 74). In 1837, an SPCA essay competition addressed the religious basis for human obligation towards animals, echoing the animal friendly discourse of eighteenth-century nonconformity (Adam et al. 2019, 7). Indeed, Preece and Chamberlain (1993, 34) consider that the success of animal welfare legislation in Britain during the nineteenth century "stemmed directly from the humanitarian impulses which lay behind the evangelical protestantism of the period".

Although reformers used nonconformist discourse, times were changing, and its influence was waning. At a time when Christianity still remained the default belief, and when most people seem not to have protested against carnism, it is reasonable to see Christian complicity in the growth of meat-eating, despite any nonconformist heritage. However, in 1822, Richard Martin claimed in Parliament, and without contradiction, that "there was not a pulpit in London that had not spoken in a pronounced manner in approbation" of his bill to protect cattle. This suggests that, in the early nineteenth century, the nonconformist heritage still continued to provide a resource for clergy to preach against cruelty, even if their congregations were more indifferent. However, the nonconformist moral sphere which made these resources relevant was shrinking.

The nonconformists themselves were accommodating to carnism, despite some evidently feeling sufficiently uneasy about it to experiment with vegetarian diets. They enjoyed the new carnist pleasures while rationalising cognitive dissonances which we may suppose especially acute for advocates of compassion and mercy. For those nonconformists who wished to enjoy the pleasures of carnism but were unwilling to accommodate to its rationalisation, the challenge was to construct a biblical warrant for it, using the resources of their discourse of meat-eating. As we shall see in the next chapter, this proved impossible without changing the very nature of nonconformist discourse itself.

References

Adam, Margaret B., David L. Clough, and David Grumett. 2019. A Christian Case for Farmed Animal Welfare. *Animals* 9, (1116): 13.

Adams, Thomas. 1862 [1629]. *The Workes*, vol. 3. London: Nisbet.
Andrewes, Lancelot. 1650. *The Pattern of Catechistical Doctrine at Large....* London: Roger Norton.
Anon. 1662. *Rump or an Exact Collection of the Choycest Poems and Songs Relating to the Late Times.* London: Henry Brome.
Baxter, Richard. 1678. *A Christian Directory...*, vol. 1. London: Nevil Simmons.
Baxter, Richard. 1691. The Poor Husbandman's Advocate to Rich Racking Landlords. In *The Reverend Richard Baxter's Last Treatise*, ed. F. J. Powicke, 1926, 39. Manchester University Press.
Bonar, Andrew. 1966 [1846]. *A Commentary on Leviticus*, 27. Edinburgh: Banner of Truth.
Borowski, Oded. 1998. *Every Living Thing: Daily Use of Animals in Ancient Israel*. London: Altamira.
Bouwsma, William J. 1988. *John Calvin.* Oxford: OUP.
Burgess, Anthony. 1652. *Spiritual Refining....* London: Thomas Underhill.
Burnett, John. 2004. *England Eats Out: A Social History 1830 to the Present.* London: Routledge.
Calvin, John. 1845 [1563]. *Commentary on Psalms.* Edinburgh: CTS.
Calvin, John. 1849 [1565]. *Commentary on Ezekiel*, 554 Edinburgh: CTS.
Calvin, John. 1965 [1578]. *Genesis.* London: Banner of Truth.
Calvin, John. [2018]. *Harmony of the Law*, vol. 2, Lev. 3.16f. Woodstock, Canada: Devoted Publishing.
Clarke, Adam. 1831. *Commentary on the Bible*, vol. 1. New York: J. Emory and B. Waugh.
Cowper, William. 2013 [1785]. *The Task and Select Other Poems*, ed. James Sambrook. London: Routledge.
Dickson, David. 1965 [1655]. *The Psalms.* London. Banner of Truth.
Dod, John and Robert Cleaver. 1635. *A Plain and Familiar Exposition of the Ten Commandments.* London: Thomas Mann.
Douglas, Mary. 1999. *Leviticus as Literature.* Oxford: OUP.
Eisnitz, Gail A. 2007. *Slaughterhouse.* New York: Prometheus Books.
Foer, Jonathan Safran. 2009. *Eating Animals.* Hamish Hamilton.
Genoways, Ted. 2015. *The Chain.* London: Harper Collins.
Gill, John. 1811 [1763]. *Exposition of the Entire Bible*, vol. 1. William Woodward.
Goodman, Michael K., and Colin Sage, eds. 2014. *Food Transgressions: Making Sense of Contemporary Food Politics.* London: Routledge.
Henry, Matthew. 1710. *An Exposition of the Old and New Testaments.* London: Nisbet.
Hollihead, Garry. 2007. Vegetarians Miss Out on the Most Exciting Textures, Flavours and Aromas. *The Independent* 18 (12): 2007.
Jamieson, Robert, A. R. Fausset and David Brown. 1871. *Commentary Critical and Explanatory on the Whole Bible.* Hartford: S.S. Scranton & Co.

Joy, Melanie. 2009. *Why We Love Dogs, Eat Pigs, and Wear Cows.* Conari Press.
Joy, Melanie and Jens Tuider. 2016. The Rise of veganism. In *Critical Perspectives on Veganism*, ed. Jodey Castricano and Rasmus R. Simonsen, v–xv. Palgrave Macmillan.
Linzey, Andrew. 2000. *Animal Gospel.* Westminster: John Knox Press.
Luther, Martin. 1960 [1538]. *Works*, ed. Jaroslav Pelikan. Saint Louis: Concordia.
Maynard, John. 1688. *The Beauty and Order of the Creation*.... London: William Gearing.
Milton, John. 1667. *Paradise Lost.* London: Peter Parker.
More, Hannah. 1840. History of Tom White, the Postboy. In *The Complete Works of Hannah More*, vol. 1, 224–231. New York: Harper & Brothers.
Nocella, Anthony J., Colin Salter, and Judy K. C. Bentley, eds. 2015. *Animals and War: Confronting the Military-Animal Industrial Complex.* Lanham: Lexington Books.
Oswald, John. 1791. *The cry of Nature or an Appeal to Mercy and to Justice, on Behalf of the Persecuted Animals.* London: J Johnson.
Perkins, William. 1602. *Cases of Conscience.* Cambridge: John Legat, book 3.
Pinker, Steven. 2012. *Better Angels of Our Nature.* Penguin.
Plutarch. 1962. The Eating of Flesh. In *Moralia,* book 3, tract 1. Harvard University Press.
Pope, Alexander. 1713. Against Barbarity to Animals. *The Guardian.* London. 61, 261–267; May 21, 1713.
Preece, Rod, and Lorna Chamberlain. 1993. *Animal Welfare and Human Values.* Wilfrid Laurier University Press.
Primatt, Humphrey. 1776. *The Duty of Mercy. Fontwell.* Susses: Centaur Press.
Rawlinson, John. 1612. *Mercy to a Beast.* Oxford: Joseph Barnes.
Ray, John. 1995 [1686]. *Historia Plantarum Generalis.* Tx. Elizabeth Mary Lazenby. Thesis no. DX185996. Newcastle University Library.
Salter, Colin. 2015. Introducing the Military-animal Industrial Complex. In Nocella et al. 2015, 1–18.
Sampson, Philip J. 2018. *Animal Ethics and the Nonconformist Conscience.* Palgrave Macmillan.
Seidenberg, David Mevorach. 2019. Veganism and Covenantalism. In *Jewish Veganism and Vegetarianism*, ed. Jacob Ari Labendz and Shmuly Yanklowitz, 161–194. New York: Suny Press.
Seneca. 1932. To Helvia, On Consolation. In *Moral Essays*, vol 2, book 12. Loeb Classic Library Harvard University Press.
Shelley, Percy Bysshe. 1947. *A Vindication of Natural Diet.* London: C W Daniel.

Stevenson, Richard J. et al. 2020. Hippocampal-Dependent Appetitive Control is Impaired by Experimental Exposure to a Western-Style Diet. Royal Society Open Science 7: 191338.

Stone, Ken. 2018. *Reading the Hebrew Bible with Animal Studies*. Stanford: Stanford University Press.

Thomas Aquinas. 1947. *Summa Theologica*, trans. Fathers of the English Dominican Province. Notre dame: Ave Maria Press, Second part of second part Q 148 Article III.

Trapp, John. 1867 [1662]. *Commentary on the Old and New Testaments*, vol. 1. London: Dickinson.

Walker, George. 1641 *The History of the Creation….* London: John Bartle.

Wesley, John. 1872 [1771]. The General Deliverance. In *Sermons on Several Occasions*, vol. II, Sermon 60.

John Wesley. 1765. *Notes on the Bible*. Grand Rapids: Christian Classics.

Wilson, Thomas. 1614. *A Commentarie Upon the Most Divine Epistle of St Paul to the Romanes* London: W Jaggard.

Yates, John. 1622. *A Modell of Divinitie*. London: Fulke Clifton.

Žižek, Slavoj. 2022. *Surplus-Enjoyment*. London: Bloomsbury Academic.

CHAPTER 10

Inventing Carno-Theology

The last chapter developed the nonconformist discourse of eating animals within the constraints imposed by subsistence farming. I argued that it subverted the dominant discourse of flesh-eating at a number of crucial points. In particular, cruelty was condemned, and the "pleasures" of meat-eating were tainted by sin. By the mid-nineteenth century, patterns of flesh-eating in England were changing as a result of urbanisation, and the industrialisation of meat production, moving farther and farther way from the biblical narrative. Some nonconformists attempted to address this, but were ultimate unable to meet the challenges faced. In this chapter, I will trace the collapse of the nonconformist discourse of eating, and its transformation into a novel way of speaking using "proof-texts" drawn from the bible.

P. J. Sampson, *Nonconformist Perspectives on Animals and Language*, The Palgrave Macmillan Animal Ethics Series,
https://doi.org/10.1007/978-3-032-12158-5_10

Citing Biblical Texts

In the U.K. and the U.S., it has now become commonplace to find both critics and advocates of vegetarianism, Christian and secular, citing biblical texts and claiming that they *oppose* a plant based diet.[1] Of course, advocates of vegetarianism deplore such perceived religious opposition, and reject its ethical claim; while critics of vegetarianism commend it, and affirm its ethical force. In particular, a specific collection of biblical "proof-texts" repeatedly occurs. However, as the term "vegetarianism" only came into general use after 1847, it nowhere appears in the Authorised Version of the Bible (1611), or those modern versions based upon it.[2] The closest phrase is the Apostle Paul's reference to those who "eateth herbs" (Rom. 14.2), but as we shall see this meant something very different from "vegetarian" when it was penned. Consequently, the proof-texts in question cannot address the issue of vegetarianism directly, so they usually adopt an indirect strategy via the antonymic pole of meat-eating. Most commonly, these proof-texts are understood to provide Biblical justifications for carnism, or even to command the practice, regardless of the radical differences between the meat-eating practices of the Ancient Near East, and carnist practices of today. A fortiori, the same proof-texts also oppose vegetarianism. I shall term the discourses surrounding these proof-texts and their exegesis in the context of diet as "carno-theology".[3]

Before the later nineteenth century, nonconformists mostly read these same pericopes in ways unrelated to the ethics of meat-eating. The transition from this nonconformist discourse to carno-theology occurred over the same period that meat production became industrialised, and more available to the general public, most of whom possessed at least a nominal Christian faith. Many Christians, both then and now, know, or at least suspect, that meat-eating is ethically compromised, yet continue the practice with relish. Carno-theology helps reduce the cognitive dissonance thus experienced, and reconfigures as biblical the addictive mechanisms

[1] For simplicity, I here focus upon vegetarianism, but similar considerations arise in debates about environmentalism and veganism.

[2] So far as I am aware it appears in no mainstream english translation.

[3] "Carno-theology" refers to a body of discourse which often employs a theological language, but it does not necessarily constitutes a "theology" in the disciplinary sense. The term draws upon Joy's (2009) term *carnism* and Derrida's (1991) concept of *carno-phallogocentrism*.

which drive it. The religious nature of carno-theology authorises practices which might otherwise appear abhorrent to the ordinarily compassionate person.

I examine the discourse of carno-theology in two respects: its use by Christian communities, especially by an anthropocentric evangelicalism which claims a nonconformist heritage, to justify a carnism which might otherwise seem deeply problematic; and its secular use to *oppose* meat-eating by associating it with an "obviously flawed" religious ideology of dominion. The once alternative nonconformist discourse, now largely subjugated, provides a point of resistance from which to survey both these carno-theological narratives.

The Invention of Meat

There is a long tradition of Christian discussion of diet based upon biblical texts, including the nonconformist discourse discussed in the previous chapter. However, these were not principally ethical discourses about meat-eating as such, let alone about carnism; the *ethical* aspect came to the fore only after the rise of vegetarianism. The earlier tradition concerned a variety of vexed theological questions including: the sinfulness of gluttony; and whether merit could be obtained by legal taboos on certain foods, or by ascetic practices (Newell 2024). The latter was often indexed by terms such as "vegetable diet" and "Pythagoreanism". For example, Hodges (1675, 41) says that those "who refuse altogether to eat the flesh of living Creatures, when duly kill'd, deserve the name of *Pythagoreans…*", not Christians.

Vegetarianism itself was a complex movement drawing upon theological, health, social, and economic factors, but its ethical dimension as "virtuous and morally responsible" has steadily gained ground (Stuart 2006. Croney and Swanson 2023, 61). The language of carno-theology grew in response to this ethical challenge, and now provides a resource to normalise a discourse of carnism as morally warranted. Advocates of vegetarianism soon responded to carno-theology with their own counter-arguments, characteristically utilising the same proof-texts. It is now commonplace to find carno-theological arguments in both secular and religious debate, by both advocates and opponents of vegetarianism. In this, it is a novel turn in the discussion of diet.

Carno-theology is also historically novel in that it often responds to a charge of moral *inferiority* against biblical Christianity. Before the twentieth century, the principal charge against Christianity concerned its rationality, not its morality. Indeed, Christians were accustomed to perceiving themselves as occupying the *moral* high-ground; indicting as immoral, say, sexual mores, sabbath non-observance, theatrical entertainments, or the consumption of alcohol. By the mid-nineteenth century this began to change, with critics pointing to historic biblical support for slavery, class hierarchy, and so on as moral deficits in the church. By the late twentieth century, such criticism had become commonplace, especially in gender politics. Vegetarianism played a significant role in this reversal, challenging Christianity for moral preeminence. As a result, carno-theology is deployed in two ways: either defensively by Christians to vindicate carnism as biblically warranted and therefore moral despite appearances; or by secular authors to allege that biblical arguments for such vindication are morally inferior hypocrisy. This chapter primarily examines its emergence in the Christian tradition to justify carnism, a dietary practice which might otherwise seem deeply problematic for advocates of love and mercy.

Carno-theology, then, not only marks a novel turn in the discussion of diet; it also marked a shift for christians from the moral high-ground to being defensively on the back-foot. I will examine specifically its development by those Christians who claimed a nonconformist heritage, and who sought to provide biblical authority for their practices. The nonconformist discourse identified in the last chapter is appropriated and subverted by a carno-theological narrative.

Social Change

Industrialisation and urbanisation brought a rapidly increasing demand for meat, and new practices to supply it, including driving large numbers of animals through cities for slaughter open to the public gaze. The animal industrial complex consolidated throughout the nineteenth century, and by 1910, there were some 43,000 butcher's shops in Britain, with "possibly more money invested in the meat industry than in any other British business" except finance (Otter 2020, 47). Post-war government subsidies greatly increased "animal production", and from the 1960s intensive factory farming had largely replaced traditional stock rearing. Supermarkets have since made meat "a truly everyday and ubiquitous item" (Goodman and Sage 2014, 10).

These changes entailed radical shifts from the traditional practices of meat-eating to modern carnism. Where humans and animals had previously lived in a mixed community, now humans meet animals on a plate; as pets; or, more recently, via images in wildlife programmes or cats on facebook. This shift in everyday experience dramatically changed our way of talking about what it is to be an animal, and the source of our moral understanding of them (Midgeley 1998, 112–125). It transformed the landscape in which the discourse discussed in the previous chapter had developed and made sense. Killing and eating animals was no longer uncommon, and traditional husbandry gradually ceased to exist. Urban congregations grew, and their daily experience of animals moved from farm to street and plate.

Rack Not Ribs

The new pattern of meat consumption utilised the dominant discourse of animals as property; in particular, as *edible* property whose consumption is both *natural*, and *pleasurable*. Culinary language had formerly co-existed with that of the farm. Now, in towns and cities, "beef", "mutton", and "pork" obliterated "cow", "sheep", and "pig". Consumer food vocabularies have since further disconnected the product on the supermarket shelf from the animal whose body part it was: nuggets, fingers, burgers, cutlets, hot-dogs (Dunayer 2001). For carnism, the presence of the animal, symbolically or in the flesh, is problematic.

Carnism Is Natural

For carnist discourse, an animal is fitted to be eaten; they are a composite of meaty parts. Meat is *Natural*; it is what animals are for.[4] Most carnists consider that meat-eating requires no justification; it is what any right-thinking, reasonable person would do (Žižek 1989, 15f). It needs no defence, as, unlike vegetarianism, it is the way things are, not an ethical stance. Meat is also N*ecessary* because it is part of a natural, healthy diet; it's our source of protein. It is a *Normal* part of life, above contestation, like nutrition itself, requiring no cultural sanction such as a special feast to permit it, let alone biblical authority. It is vegetarians who need to justify

[4] The capitalised terms Necessary, Natural, Normal, Nice, here re-grouped under two categories of *naturalness* and *pleasure*, are drawn from Piazza et al. (2015).

their abnormal diet, not normal people. This discourse has been very influential, and its traces can be found even among critics of the contemporary meat industry. For example, Michael Pollan (2002, 9) argues that "we humans have been eating animals as long as we have lived on this earth. ...[It] is part of our evolutionary heritage... Eating meat helped make us what we are, in a social and a biological sense. Under the pressure of the hunt, the human brain grew in size and complexity, and around the fire where the meat was cooked, human culture first flourished". Hence, he says, "meat eating is something very deep indeed". Meat, in the carnist sense, was being invented; the ethical sphere formerly delineating human interactions with animals contracted dramatically as daily contact with living animals became daily contact with dead ones on a plate.

Carnism Is Pleasure

A specific experience of pleasure corresponds to this ontology of what animals are. In carnist discourse, meat is *Nice* ("I could not live without bacon"); we experience animals as a pleasing aroma, a taste and texture in the mouth, a desire for more. As Steven Pinker (2012, 551) observes, meat "tastes good, and eating it makes people happy". The meaty parts of an animal are differentiated by edible pleasure, not function. A leg of mutton is celebrated for its succulence, not its use in walking; a chicken wing for its juicy flavour, not its suitability for flight. Indeed, in modern carnism, bodily function is often compromised for culinary pleasure: a chicken's legs break under excessive body weight; mutilation is routine. Pleasures once regarded with suspicion in nonconformist discourse as the gluttonous excesses of the rich become commonplace.

The transition from meat-eating to carnism made animal flesh desirable in new ways. In traditional animal husbandry, most people ate to live; the little flesh available was not necessarily appetising. Where meat-eating at special feasts did create a desire for more, it was not one that could often be satisfied in everyday life. Once flesh became "a truly everyday and ubiquitous item", the desire could be both created and daily satisfied. A space opened for excess desire to also become "ubiquitous", not the preserve of the rich. This was to prove deeply problematic for nonconformist discourse.

Carnist discourse is radically different from the nonconformist discussion of flesh-eating set out in the last chapter. In particular, it presents both an incompatible ontology and an irreconcilable account of human pleasure. Animals created as song to manifest and praise God are now made for human consumption. Surplus pleasure and wonder at creation now transforms into gastronomic excess.

A Welcome Dilemma?

Most people in industrialising cities welcomed carnism as progress. It brought the promise of a nutritious and pleasurable diet for all, not just the rich. Indeed, by the late nineteenth century, meat-eating had became routine. The pleasures of a succulent leg of lamb or the crisped skin of a pig became the norm for an unprecedented number of people. However, some had reservations, mainly on ethical and health grounds. In the mid-nineteenth century, unease crystallised in the vegetarian movement in England which rejected eating animals outright, arguing that carnism was neither *natural*, nor were its *pleasures* ethical.

Nonconformists faced a dilemma. The kind of animal husbandry assumed in the bible and practised in early modern England was disappearing, and with it the moral sphere which nonconformist discourse had created. That discourse had been honed to the daily life and concerns of farming communities, but preachers now addressed urban congregations who were becoming regular meat-eaters, and rarely visited farms. Preacher and congregant alike were parting company with the experience of animal husbandry represented in the biblical narrative; fewer sermons about animals bred less familiarity with nonconformist discourse among parishioners. The nonconformist discourse of animals became both less relevant and less familiar to congregations. The eighteenth-century congregationalist, Jonathan Edwards, lamented that men were not uneasy that many thousand of animals "every day cease to live, and so have an end put to all their pleasure, at butchers' shambles in great cities" (Lane 2011, 121). But even if they *had* been uneasy, they simply didn't know what to do about it; nonconformist discourse, honed to address the cruel individual, offered few resources in an age when cruelty to animals was becoming part of the texture of life.

Worse, nonconformist ambivalence towards meat-eating could look suspiciously like a fashionable dietary regimen. John Wesley (1872 [1771], para. 1) had adopted a vegetable diet in about 1735, probably on medical advice, but certainly strongly resonating with his preaching that God "directs us to be tender of even the meaner creatures; to show mercy to these also". The Bishop of London took him to task over his diet, claiming that Methodists liked to give the appearance of uncommon sanctity by abstaining from animal food. Wesley (1931 [1747], para. 14) responded, saying that he had abandoned his vegetable diet for the sake of some who thought he "made it a point of conscience". Wesley cited biblical warrant for his decision, accurately fitted to the accusation about abstinence from meat bringing sanctity (1 Cor. 8.13). For nonconformists like Wesley, a failure to conform to the conventional carnist diet was already providing an occasion for criticism by establishment Churchmen who embraced the dominant discourse.

From the turn of the nineteenth century, some nonconformists did attempt to tailor their discursive resources to meet the new circumstances. Rather than preach to urban congregations largely uninvolved in animal husbandry as individuals, they began applying their discourse to law reform, and to animal welfare societies such as the SPCA. As noted in the last chapter, there were successes, but the influence of nonconformist discourse decreased throughout the century, at the same time as the pressure to conform to carnism increased. Two factors spurred this decline and ultimately contributed to ending nonconformist resources for animal advocacy: the rise of an ethical discourse pioneered by the vegetarian movement; and a change within nonconformity itself. I will deal briefly with each in turn.

The Subjugation of Nonconformist Discourse

The Birth of Vegetarianism

As we saw, Wesley's adoption of a vegetable diet was criticised as an ostentatiously sanctimonious departure from the dominant norm of carnism. By 1847, when the Vegetarian Society formed, the tone of the discourse had shifted, and secular ethical debate steadily gained ground over theological niceties. Although vegetarians could still be dismissed as eccentrics, their criticisms touched a nerve with christians, not least because they were claiming the moral high-ground, formerly the preserve of Christianity. This novel challenge to their probity generated a renewed discussion of diet, especially among those who sought guidance in the bible. In particular, they tried to defend carnism against vegetarian charges of immorality, gluttony, and even "Saturnalian" excess "under the garb of Christian sentiment" (Salt 1914, 32).

The Birth of Evangelicalism

The birth of vegetarianism, highlighting both cruelty and dissolute pleasure, coincided with the waning of nonconformist discourse under the impact of a new *evangelicalism,* with a novel social imaginary: an embrace of values, institutions, and symbols preoccupied with anthropocentric emotional expression. Although it saw itself as the heir to nonconformity, evangelicalism differed in three respects which will be important for our discussion. Its anthropocentric expressionism, foregrounding individual human experience, displaced the theocentrism of nonconformity; its expression of the world as a vale of tears, destined to be burned up, displaced a creation manifesting and praising God; and the narrowed focus of its "biblicism", emphasising isolated pericopes as transparently accessible propositions, displaced contextual reading (Bebbington 1989). This came to a symbolic head when C. H. Spurgeon (1887) protested at the "downgrading" of the "evangelical orthodoxy" of nonconformity (Swanson 2001. Lloyd-Jones 1992, 12f).

Theocentric nonconformist discourse continued to wane into the twentieth century, and congregations were rarely reminded that animals are created to sing praise to God, not as ingredients.[5] Of course, this was a gradual process, which left many remnants of an earlier language. Vestiges of nonconformist discourse often survived in hymns calling upon animals to praise God, although these were now given an anthropocentric and expressive sheen. Thus congregations continued to sing Thomas Ken's seventeenth-century exhortation: "Praise God from whom all blessings flow; praise Him, *all creatureshere below*"; or Isaac Watts early eighteenth century "Let *every creature* rise and bring peculiar honours to our King". Nonconformist discourse read these as robust theocentric calls for animals to fulfil their purpose in singing praise. Appropriated by evangelicalism, the hymns of Ken or Watts became essentially expressive metaphor; pious thoughts that were left in the pew at the end of the service. Congregants could figuratively exhort animals to sing praise to God, while then going home to eat the choir.

[5] I first heard the phrase "animals are ingredients" from Andy Hamilton on BBC Radio 4 in June 2004.

The End of Nonconformist Resistance

The industrialisation of meat, the ethical challenge from vegetarianism, and the rise of an anthropocentric and expressive evangelicalism, all combined to effectively draw the teeth of the most effective resistance to the dominant discourse of animals that then existed. However, discussion of carnism did not altogether cease among Christians, especially within evangelicalism.

By the mid-nineteenth century, the everyday human experience of animals in cities had moved from farm to plate. Most Christians were now tacitly conformed to carnism, routinely eating meat which they found both natural and pleasurable. It was not entirely possible to ignore the publicly visible cruelty, but this was just the ways things are, something individuals are powerlessness to change. This comfortable acquiescence in carnism was disturbed by vegetarians who brought moral scruples, traditionally the strong point of Christians.

Vegetarians believed there was another, better way to live, and their ethical rebuke stung. The expressive emphasis of evangelicalism made the challenge especially acute. In an industrial age, Christians needed a new language of meat-eating as both *natural* and *moral*, one which would deal with the cognitive dissonance provoked by both vegetarians, and their own consciences.

Nonconformist discourse, formed in an earlier age, was not up to this task; indeed, it had done the opposite to what was required. It had *de-naturalised* meat-eating by denying it a creational mandate. Moreover, it had *morally tainted* meat-eating with the sins of both cruelty and transgressive pleasure. I am not aware of any defence of carnism drawing upon the nonconformist resources set out in the previous chapter.

The Certainty of the Bible Text

Something far more robust was required to justify an activity that might otherwise appear deeply problematic. A carno-theology was needed to provide biblical warrant and moral certainty, preferably a specific command to eat meat. For Christians who inherited the nonconformist respect for biblical authority but not its conclusions, this meant reading the bible anew (Lloyd-Jones 1992, 35, 42). Then carnism and its pleasures could be enjoyed with a clear conscience.

Inventing a carno-theology was no small task. Meat-eating has to go from being tainted by sin, to being authorised by God; from an ambivalent necessity to a good in its own right; from a transgressive excess to normal pleasure of life. Spurred by this challenge, evangelicalism re-read the two intersecting sub-discourses of ontology and pleasure which had founded the nonconformist discourse of animals.

Firstly, ontologically, meat must become natural; animals must be authorised by God as created edible.

Secondly, meat-eating must be normalised as Nice. Its pleasures and desires must appear normal, certainly not tainted by sin and excess. Lest it dampen carnist pleasures, animal cruelty must be either justified, insignificant, or regrettable but necessary.

Such a discourse entailed a major transformation of that of nonconformist meat-eating described in the last chapter. That had been a complex interconnected network of threads woven into the everyday life of both the biblical narrative and the nonconformists' own experience. It encompassed both the vegetable diet of Eden and that of the restored world; the necessity of meat-eating as a result of sin; the wickedness of cruelty; the restriction of excess pleasure; and the impossibility of gaining spiritual merit through either legalistic or ascetic patterns of eating. By contrast, evangelicalism's carnist discourse need not be concerned with such matters. It was only required to show directly from the bible that meat is authorised as food, even when industrially produced. To show that, as Derek Antrobus (1997, 16) puts it, the bible "is a meat-eater's manifesto". Enter the proof text.

Carno-theology and Proof-texts

As we have seen, nonconformist literature is often thick with biblical citations, either intertexted or made explicit. These were referred to as "scripture *proofs*", "*pregnant texts*", or "*proving*" a truth. For modern readers, this suggests an argument compelling acceptance; "proof" nowadays entails certainty. But before the nineteenth century, a "proof" was more ambiguous, carrying also the sense of a "test" or a "trial". This now survives only in phrases such as "the exception that *proves* the rule", and "the *proof* of the pudding". Richard Muller (2003, 511; also 417, 450) suggests that early modern usage indicated, not a set of verses providing summary certainty, but "a sign that reading of scripture has contributed" to the argument and "is recommended" to its readers. Nonconformist "scripture proofs" were not isolated verses. They were set within the narrative arc of the bible, and could be very extensive, sometimes of book length. Thus Anthony Burgess' *Treatise of original sin... **proving**...it ... by pregnant texts of Scripture* ran to more than 550 pages; while Richard Baxter's *Plain Scripture-**proof** of Infants Church-membership...* restricted itself to a mere 350.

Nonconformist discourse also regarded the biblical text as "perspicuous", not requiring a complex system of interpretation such as the four-fold method of medieval scholasticism. The meaning of the text is, as Luther (nd: 1525, 14–15) puts it, "manifested openly", although our understanding of it cannot be divorced from the work of the Holy Spirit. The Westminster Confession (1647, 1.vii, ix) says that anyone may understand the biblical text by "due use of the ordinary means" of study. This does not mean that we may "sleepily" look over the text. For example, because not everything is equally plain, its meaning "must be searched and known by other places that speak more clearly". Phrases such as "due use" and "searched" indicate careful study, giving attention to context, genre, and other relevant aspects, often in the original languages. Evangelicalism appropriated both "scripture proof" and "perspicuity", and transformed them both.

First it appropriated nonconformist "perspicuity", and transformed it into the certainty of a transparently accessible text, often a single verse in isolation. This assumes, as Kevin Vanhoozer (2005, 271) argues, "a uniform propositional revelation spread evenly throughout Scripture" which relieves the reader of any need to consider the context of either its production, setting, or reception. Texts acquire a sheen of industrial uniformity, unencumbered by context.

Secondly, the nonconformist "scripture proof" was appropriated as the "proof-text", and transformed to eliminate the sense of testing, leaving only scientific certainty as an alibi against the charge of immorality brought by the vegetarian critique. Nonconformist "scripture proofs" *resisted* accepted cultural norms of cruelty. In carno-theology, this is transformed into proof-texts oriented around *defending* accepted cultural norms of carnism. Within a community which values biblical proof-texts, this is not a matter of abstract interpretation but of practical significance. Proof-texts provide a framework for behaviour; indeed, sometimes for perverse, even sinister, manipulation of behaviour.[6] Evangelicalism is particularly reliant upon the proof-text where an issue is not well integrated into its religious tradition. While proof-texts are very common when it is a question of a BBQ steak or a bacon sandwich, they may co-exist with a more nuanced discussion of other topics, such as the doctrine of the Trinity. Muller (2003, 509), contrasting the modern proof-text with seventeenth-century practice, calls them "the rank citation of texts apart from their context and apart from any consideration of the results of exegesis [of the text in question]". As Gordon Fee (1993, 18) observes of proof-texts, "the bible is not a collection of propositions, of 'Sayings from Chairman God'".

Ethical Certainty

Far from Burgess' or Baxter's tomes, proof-texts are shortened from book length treatments of a subject to routinised, isolated verses; they provide a resource for ethical certainty about a contested issue, not a space for ethical enquiry. The proof-texts of carno-theology are selected and organised around normalising the accepted cultural norms of carnism. Formerly disparate collections of pericopes are unified as a coherent ensemble, engineered by carnism itself, with its own internal structure and dynamics. Each pericope is read in a new way, foregrounding relationships between them not previously emphasised, if noted at all.

In engineering such an ensemble, the excluded or marginalised is as significant as what is included. Keith Thomas (1983, 24) documented examples of christian anthropocentrism in the seventeenth century, noting that they "tended to disregard those sections of the Old Testament which suggest that man has a duty to act responsibly towards God's creation". Evangelicalism took up this mantle. Margaret Adams (2019, 3) notes the reliance of carno-theology on "a few select biblical passages, while neglecting others". Through the proof-text, the animal industrial complex extended its butchering power from the slaughterhouse into the religious institution of evangelicalism, chopping apart nonconformist discourse and

[6] For example, Gen. 9.18–27 and Eph. 6.5–7 were prominent in authorising slavery. According to witnesses to the Makin review of allegations of abuse against John Smyth, he cited as proof-texts 2 Sam. 7.14 and Heb.12.4 "to assist his abuse" (Makin 2024, 8.6.5.a, 11 March 12).

packaging it into a conveniently unified ensemble of texts. Carnism authorised what was included and excluded, generated novel connections, and sidelined old ones.

Just as the nonconformists drew upon the Renaissance scholarship of their day to cut through the complexities of Thomism, so evangelicalism drew upon the language of science, technology, and industrial society to create its own way of reading the bible. Vanhoozer (2005, 270–1) notes the likeness between the proof-text of evangelicalism and the propositional statements of positive science. Evangelicalism looked through the lens of industrial, even consumer, society, customising the biblical text into the Maker's instructions, a "handbook for life", as though life were a piece of consumer technology (Bouma 2015. Million Voices 2024). However, where nonconformists had recognised their own culture in the animal husbandry of the biblical texts, the world of evangelicalism was entirely foreign to those texts. Neither positive science nor industrialisation were features of the Ancient Near East; there was no "handbook" or instruction manual for a camel. The same industrial processes which had generated carnism and its meat products, also fashioned carno-theology and its proof-texts.[7]

Carno-theology and Vegetarianism

Carno-theology developed hand in hand with vegetarianism as both responded to the growing animal industrial complex. The cruelties entailed were often publicly visible, certainly widely known, and such suffering naturally induced revulsion in all but the most de-sensitised. Adherents of the new evangelicalism commonly regarded themselves as both sensitive and morally aware. To collude in seemingly bad things is repugnant, and requires a higher authority to show that they are not really bad, and to anaesthetise the soul against the suffering caused. Evangelicalism's understanding of "tough love" illustrates the mechanism. Christians should love their children, but traditional childrearing practices include physical beating, which might seem unloving. Appeals to proof-texts justify what might otherwise seem wrong or even perverse (e.g. Prov. 22.15, 23.13, 29.15; see Makin 2024, 8.6.5.a, 11.3.12). The greater the

[7] I am indebted to Rev. Dr. Timothy Hull for the example of the camel and her handbook.

evil in which we collude, the greater the authority required. For evangelicalism, there is no higher authority than the bible. To indulge the egregious violence of carnism required nothing less than biblical warrant. But the human soul is always altered by violence (Žižek 2009).

As a religious language authorising carnism, carno-theology was especially congenial to consumers in the late nineteenth century when most had at least a nominal christian faith; perhaps without such religious justifications, many ordinarily compassionate people would not have found it in their heart to sanction such cruelty. To this day, especially in the U.S., carno-theology remains prominent in debates about meat-eating. It provides evangelicalism with a divine authority for what might otherwise be recognised as wickedness.

Naturally, such discourse was less congenial to vegetarians, and exchanges about proof-texts became common by the turn of the twentieth century. A Vegetarian Society tract of 1914 imagines carno-theology's response to an ethical challenge:

"The question is, to me, a religious one. The animals were sent us as food, and if Peter was bidden to arise, kill, and eat, it would be impious in us to abstain... Besides... what does it matter what we eat? It is not that which goeth into the mouth that defileth the man, but that which [comes from the heart]" (Salt 1914, 32).

This intertexts, without reference, two biblical proof-texts (Acts 10.13, and Matt. 15.10), taking it for granted that readers in 1914 would be familiar with their biblical provenance and relevance to the debate. The tract advances counter-arguments, and concludes that this supposed Christian justification is a hypocritical defence of excess in Christian guise.

The biblical allusions in the Vegetarian Society tract are representative of two ensembles of proof-texts, challenging the two dimensions of nonconformist discourse: ontological status, and pleasure. Directly contrary to nonconformity, evangelicalism asserts that meat is naturally food; and that the bible authorises its pleasures. These two dimensions are often referenced by iconic terms such as "dominion" (Gen. 1.26–8), or "Peter's dream" (Acts 10.9–16), evoking all of carno-theology in one or two words.[8] I will discuss these two ensembles of texts before exploring further their use in religious and secular discourses. Space

[8] Sometimes a single word can suffice. The popular worship leader Sean Feucht (2023) calculates that there are 274 bible verses that mention 'meat', but only 3 that mention vegetables. Therefore, he says, "be biblical eater and consume meat". He does not

prevents a discussion of less common pericopes, or of those omitted from carno-theology because they are less congenial to an anthropocentric appropriation. It is not my purpose to provide exegeses, but simply to indicate how nonconformist discourse has been appropriated and transformed.

Ontological Authorisation

This ensemble of proof-texts aims to show that God authorises men to eat meat by establishing that animals were created as food. It is often set within a broader anthropocentric discourse that "animals were intended for man's enjoyment and use" (Anon 2025a, b). As Sarah Palin (2009, 133), a self-identified bible-believing "evangelical" christian, colloquially put it, "If God had not intended for us to eat animals, how come He made them out of meat?". To prove its ontological case, carno-theology must show that animals are a general singular, a species, made for men to eat, made of meat; they are fit to die. By contrast, humans are individuals with a right to life, made to eat flesh; consumers, not conductors. Finding a biblical warrant for this manifesto faces immediate difficulties.

Animals Aren't Individuals

For carno-theology, animal "species matter more than individuals". "Clearly God treats animals as groups, but people as individuals and groups"; consequently, individual animals do not have a right to life, and we may kill them (Vantassel 2009, 169).

Yet, according to Jesus of Nazareth, God cares, even for small birds, *individually*, not simply as a species in the general singular (Matt. 10; Lk. 12). The parable of the lost sheep only makes sense if the shepherd (representing God) cared for her sheep as individuals (Lk. 15.4f). Similarly, the cattle of Nineveh are not protected to preserve a species but as individuals, like the many people also mentioned (Jonah 4.11); the ewe lamb in 2 Samual 12 is very much an individual; Balaam was preserved on account of the individual animal he rode (Num.22); and Jesus compares himself with an individual hen with her chicks to illustrate his own individual concern for others (Matt 23.37). More generally, nonconformist

specify which translation he is using. I assume this is a joke, parodying evangelicalism's hermeneutics.

discourse is clear that God's providential care extends to individual creatures. As Calvin (1958, VI.266) notes: "Christ defines the providence of God very differently from those who, ... imagine the workings of providence in a confused way, as though God paid no attention to individual creatures. Christ, on the other hand, declares that every single one of God's creatures is under his hand and care, and that nothing happens by chance".

Reality and Metaphor

Secondly, as we saw in Chapter 2, biblical texts speak of animals as created to manifest and praise God. Consider the celebration of life in Psalm 148.7f which exemplifies such a nonconformist ontology:

> Praise the Lord from the earth, ye dragons, and all deeps: fire, ... mountains, and all hills; fruitful trees, and all cedars: beasts, and all cattle; creeping things, and flying fowl.. praise the Lord.

For carnists, the psalm would have to read a little differently:

> Praise the Lord from the earth, ye exotic meats, and all deeps: fire, ... mountains, and all hills; meat, and all beef steaks; foie gras, and roast chicken breasts, all slaughtered things, ... Praise the Lord.

Most people who think of christianity as affirming life would recognise that something is wrong here.

Carno-theology avoids this potential difficulty by sharply distinguishing between *real animals* (created for men to eat), and *metaphors* which speak of praising and manifesting God. According to carno-theology, when the bible says that animals *praise* God, we should not take this literally; it is only a metaphor. We might speak of a lamb praising God when gambolling in the fields; but this is only figurative, since she is really designed to be butchered into chops. Moreover, animals do not *manifest* God, only humans do that. "God created adam alone–both Adam and Eve–in His Own Image. He did not create animals in His Image" (Bayly and Tim 2011). This strategy enables congregations to remain comfortable with uplifting animal *metaphors* while eating the very *literal* flesh. Interestingly, the assertion that the bible cannot be taken "literally" is unusual in evangelicalism which generally prefers "literal" interpretations.

Carno-theology not only seeks to establish the negative case that animals are not individuals and that the *manifestation* and *praise* of God are only metaphors, it also seeks to make the positive case that animals are fit to be eaten through a discussion of dominion.

Dominion

In carno-theology, "dominion" is an iconic term summarising the animal's ontological status as created to be eaten (eg Wilkinson 1990). It is sometimes termed "dominionism", pointing to humanity's "superior status in creation, and that this status provides a moral basis for humans to wield power over nature, and to compel it to serve human needs and interests". Dominionism claims that this status is ontologically established in Gen. 1.26–8, and includes "authority over an animal's life and death" (Vantassel 2009, 2–3).

The proof text cited by Vantassel narrates that humans were created "in the image of God" and given "dominion over the fish of the sea, and over the fowl of the air, and over the cattle, and over all the earth, and over every creeping thing that creepeth upon the earth". The very next verse authorises a vegetable diet, with obvious implications for the meaning of "dominion"; Vantassel is in the majority of carno-theologians in omitting this.

Often, simply citing this text authorises carnism. Thus, the then Minister of Agriculture and sometime member of The Church of England General Synod, John Selwyn Gummer (1990), observed that "The Bible tells us that we are masters of the fowls of the air and the beasts of the field and we very properly eat them". The then Minister referred to vegetarianism as "wholly unnatural". In such readings, the provision of a vegetable diet in Gen. 1.29, human sin, and the Fall are all sidelined. The appropriation of this pericope by carno-theology is similar to that of dominion by the dominant discourse which I discussed in Chapter 1, and which I will not repeat it here.

Some carno-theology recognises that, since Eden's "dominion" entailed a vegetable diet, animals weren't made to be eaten, and God didn't authorise meat-eating. A tactical turn to the Flood narrative attempts to retrieve the situation (Gen. 9.3): "the strictly vegetarian program outlined in Genesis was lifted immediately following the Flood" (Anon 2025b). Another author tells us that the "Bible is clear that God gave humans permission to eat meat. In Genesis 9:3 … God gave humans permission to eat meat, and it was part of the original diet given to us"

(Anon 2025c). The actual vegetable "original diet" of Eden, and the sin which occasioned this concession, are elided, as is meat's consequent taint.

Indirect Carnism

A subset of proof-texts approach the ontological status of meat indirectly, through its opposite, vegetarianism, which is presented as a "weakness" of faith, contrary to the command to "kill and eat", or even as "demonic" (Acts 10.9–16, Rom. 14.2, 1 Cor. 8.4–13, 1Tim 4.1–3). These proof-texts are drawn from pericopes dealing with a variety of vexed issues in early Christianity, including whether it is possible to gain spiritual merit through either an ascetic or legalistic diet; whether, by eating food sacrificed to idols, the eater imbibes the spirit of the idol; and whether the hitherto Jewish gospel should be preached to gentiles (Newell 2024). Carno-theology repurposes these verses by chopping them from their context. The proof-texts are now read as condemning vegetarianism, thereby promoting carnism.

Carno-theology's appropriation of a pericope in Paul's letter to Timothy is of particular interest. Addressing a group who were "speaking lies in hypocrisy" by forbidding marriage and commanding abstinence "from meats", Paul refers to "doctrines of devils" (1 Tim 4.1–3). Carno-theology reads this as a proof-text that vegetarianism is demonic.[9] The Authorised Version accurately renders as "meats" the greek which meant "food" generally, as did the English "meats" in 1611; carno-theology contentiously identifies "meats" (food) with "meat", the flesh of an animal. Forbidding marriage and food suggests that Paul was referring to an extreme form of first-century asceticism, not a feature of twenty-first-century vegetarianism (Newell 2024).

The second ensemble of proof-texts deployed in carno-theology concerns the authorisation of carnist pleasure.

[9] This seems so unusual, that I give several examples. Ditzel (2011) and Anon (2025d) concentrate on the link between vegetarianism and the demonic. Bayly and Tim (2011) focusses on veganism, claiming it is "a Satanic conspiracy against God's Creation Order". Sometimes the argument extends to the animal rights movement as a whole (Matherson 2015).

Pleasure

Carnist pleasure raises two issues: the eater's pleasure in eating; and the reliance of this pleasure upon the suffering of the eaten.

Nonconformist discourse was deeply ambivalent about the excess pleasures of meat-eating, and condemned pleasure derived from another's suffering outright. Carno-theology, tasked with justifying a carnism which might otherwise appear cruel and excess, needed something far more certain than ambivalence; it also had to address the issue of suffering. This became especially necessary from the eighteenth century, as the public cruelties of industrialising meat production were increasingly visible to compassionate people. Animal suffering to make meat was undeniable, yet carnist pleasure must be enjoyable anyway (Žižek. 1989). As the eighteenth-century literary figure, Oliver Goldsmith (1870, 191–2) remarked, "Strange contrariety of conduct! they pity, and they eat the objects of their compassion!".

This "contrariety" might be resolved at the individual level by carnist defence mechanisms and moral disengagement. But public debate needed more than this (Joy and Tuider 2016, viii–xii. Bandura et al. 1996. Graça et al. 2016). Carno-theology needed to reconfigure as biblical, both the very excess pleasure which nonconformist discourse had frowned upon; and the cruelties involved in the animal-industrial complex.

Live the Dream

Like its nonconformist precursor, evangelicalism has historically positioned itself against Roman Catholicism, especially criticising ascetic traditions which are understood as "works religion"—that is, as attempts to obtain spiritual merit by dietary restraint. In particular, evangelicalism appropriated the nonconformist rejection of the view that abstaining from meat bought merit. This already provided a general defence of excess pleasure, since abstaining from meat, and thus from its pleasures, is undesirable as it cannot bring merit. A nonconformist rejection of salvation by asceticism had been transformed into an embrace of the pleasures of carnism. By associating dietary asceticism with vegetarianism, a rejection of the former authorised a rejection of the latter (Cole 2008). However, this is a little convoluted, so it would be better if a more direct proof-text were available. Enter Peter's dream.

Peter's dream has become iconic not only of God's supposed command to eat meat, but also of authorising its pleasures. Its exegesis is beyond the scope of this discussion, and concerns the inclusion of Gentiles in the hitherto Jewish gospel (Acts 10. 9–11, 13, 15; Gal 3), but the manifest content is clear. In a vision or dream, God commands Peter to kill an animal and eat it. Peter protests, as he considers the animal contaminated ("common") by its contact with ceremonially "unclean" animals which should not be eaten; God responds by rejecting Peter's understanding of contamination. Peter comes to realise that the latent content of the dream indicates that Jews would not be contaminated by contact with Gentiles in the church. In carno-theology, this complex narrative arc is reduced to the proof-text of God's command to Peter to kill and eat "unclean" animals (such as pigs), especially emphasising the pleasures of bacon (Anon 2019). A popular cartoon depicts Peter with an aromatic slice of bacon, and is captioned "Acts 10: Live the Dream". Salvation of the Gentiles is replaced by salivation at the smell of bacon.

A Sweet Smelling Savour

An even more direct proof-text demonstrating that carnist pleasure is approved by God is generated by the phrase "sweet smelling savour" or its close cognates, which occur some forty times throughout the AV bible. It is used in relation to the sacrifice by fire of animal flesh, vegetables, and wine, although carno-theology focusses on the smell of roasting meat. It first occurs in Genesis (8.20–21) when Noah offered burnt flesh as a sacrifice, and "the Lord smelled a sweet savour", initiating the covenant discourse of Genesis, chapter 9. In carno-theology, this is widely read in isolation from its discursive setting, as though God approved the sacrifice because He liked the smell.

For modern people, as soon as fire meets flesh, men man the tongs. A google search of "people cooking BBQ" images revealed a ratio of 10 to one of men to women. Similarly a google survey showed words associated with BBQ to include: "macho", "primal", "caveman", a "nice char", and "great sear". The smell of burning flesh seems to evoke a certain kind of masculinity which harks back to experiences attributed to our earliest human ancestors. Presumably stone-age women specialised in in-cave, steamed broccoli.

It is this hegemonic masculinised representation of burning-flesh-as-BBQ which frames readings of these "sweet smelling savour" proof-texts, rather than the covenant discourse of the text. It attributes to a masculinised God a desire which nonconformist discourse had found tainted by sin.

Carno-theology reasons that, if God Himself relishes the smell of roasting flesh, surely this is proof enough that God intends it to be enjoyed, just like a BBQ. In his meditation on "God's Favorite Smell", Mark Roberts (2010) tells us that "God is especially fond of the smell of burning sacrifices. In some cases, these sacrifices would smell just like meat being grilled over coals on a summer's evening". One commentary tells us that "the Lord apparently has always liked the smell of Barbecue" (Stone 2018, 67).

This reversion to the pagan idea that a sacrifice is made to whet or satisfy the appetite of a god was roundly rejected in nonconformist discourse. "It is indeed sure that God, who breathes life into all, and borrows nothing from any, does not want food", observed Calvin (2018. Lev. 3.1–17.16).

Nonconformists translated the "sweet smelling savour" of the AV variously as "odour of rest", "quieting odour", or savour of rest, breaking with the culinary association, and placing it within its covenant setting. The image has a relish of salvation, reminding us of the terrible price in suffering which was to be paid for peace. Ainsworth (1843 [1627], Gen. 8.21) considered it a play on words: "a noahic odour", based upon the similarity between "rest" and "Noah" in Hebrew. Calvin (1965 [1578], Gen. 8.21. 1845 [1559], 2.1.7) sharply rejects the culinary association: "nothing can be more absurd than to suppose that God should have been appeased by the filthy smoke of entrails, and of flesh"; "what could be more vain or frivolous than for men to reconcile themselves to God, by offering him the foul odour produced by burning the fat of beasts?". Evidently not a big fan of barbecues.

The sharp rejection of the smell of burning flesh as an appetite stimulant for God, is reinforced by the fact that the phrase appears three times in the New Testament, echoing the septuagint greek of the hebrew passages, and all referring to Christ's crucifixion as a self-sacrifice to God "for a sweet smelling savour" (Eph. 5.2; Phil. 4.18, 2 Cor. 2.15). Plainly we are not intended to understand as an appetite stimulant the smell of the body fluids which accompanied a man being tortured to death on Golgotha Hill. That, as Thomas Goodwin (1861, 213–4) notes, would be "abominable". By displacing the relish of salvation with that of the BBQ, carno-theology here borders upon what nonconformists would have considered blasphemous.

The second aspect of pleasure discourse which carno-theology needs to address is the suffering of the animal whose flesh carnists enjoy eating.

Cruelty

Carnists experience *meat-related cognitive dissonance* when their desire for meat conflicts with their sense of not being a cruel person (Rothgerber 2020). This may be more acute within evangelicalism because of its explicitly espoused values of compassion and kindness. Carno-theology has two ways of dealing with this: silence; and through a discussion of sacrifice.

Silence

Cruelty, a major theme of nonconformist discourse, scarcely appears in carno-theology; a discursive invisibility echoing that of carnism's victims, kept out of sight and conveniently out of mind (Joy and Tuider 2016, viii). Just as every meal of meat has, as absent referent, the animal fed upon, so carno-theology absents the cruelty that the animal endured, cloaking the violence of carnism (Adams 2011). Thus Vantassel (2009, 57) is able to observe that since his concern is "the treatment of animals rather than whether they suffer, [the category of suffering] will not be discussed". Most people find that empathy forces us to recognise the suffering of others, whether humans or other animals. Consequently, if we are *not* to see suffering, if it is to be made invisible, we need to *not* empathise. Indeed, some influential evangelicals have recently argued against an "untethered" empathy which might lead us to identify too closely with the suffering of another (Rigney 2025). At the time of writing, this rather niche rejection of empathy has become more widely debated in the context of the purported remark by Charlie Kirk in 2022 that "empathy is a made-up, new age term" (Liles 2025). For legal theorist, Lisa Johnson (2017, 71–3), refusing to empathise with an animal in need of help, nurturing an indifference to suffering, is to "nurture evil". Conversely, to "help animals that ask for our help" is to enable "a state of wonderment about his or her encounter with a human steward".

Where carno-theology does mention the suffering of animals, it is expressively normalised, part of this "vale of tears", regrettable no doubt, but necessary; or else marginalised by comparison with the suffering of humans who are made "in the image of God". One website which *does* mention cruelty is clear that it "isn't consistent with a Christian world view", citing Prov.12.10: "Whoever is righteous has regard for the life of his beast". This is compatible with the nonconformist view set out in Chapter 5. However, unlike the vast majority of nonconformist texts, the author cites only the first half of the verse, omitting the second half which refers to the *wickedness* of cruelty. *Wickedness*, though, is not entirely neglected. The website continues: "many animal advocacy groups don't acknowledge that animal life shouldn't be given priority over human life. Man alone has been made 'in the image of God' (Genesis 1.27). Because of that, denying starving people food to protect the 'rights' of birds and beasts is *wicked* and foolish". No evidence is given that animal advocacy groups deny starving people food, but the simple assertion that they do so displaces *wickedness* from people who are cruel to animals onto people who are advocates for animals (Anon 2025b, my emph.). And so wickedness is banished from carno-theology, which now has only righteous men, and is lodged in those who abhor cruelty. Carno-Theology reverses the sense of Prov. 12.10, achieving its aim of making wickedness acceptable.

Sacrifice

The second strategy of carno-theology for warranting cruelty entails normalising violence for epicurean purposes through a discourse of sacrifice. To avoid seeming transgressive, violence against living beings requires an authorising agency; in carno-theology, this is the bible. As Carol Adams (n.d.) observes, "Christian notions such as dominion and sacrifice are perverted to maintain perverted relationships undergirded by violence". Specifically, carno-theology finds in the Pentateuchal law many proof-texts for God sanctioning sacrificing animals and eating their flesh. In the previous chapter, I discussed the background of the sacrificial system in practical animal husbandry, and I will not repeat it here. Carno-theology transforms laws which restrict cruelty and excess into the authorisation of arbitrary violence upon the bodies of an arbitrary number of animals. Laws which frame slaughter as a reminder of sin are transformed into that fleshy commodification of body parts which is carnism. The sin which taught violence and cruelty in the first place is sidelined.

Sidelining Sin

Evangelicalism, both within and without its community of adherents, is associated with the identification and attribution of sinful behaviours, especially those connected with sexuality. Carno-theology is also concerned with sin, but now with showing that seemingly reprehensible behaviours are *not* sinful. Its ontological discourse sidelines sin, the Fall, and the violence which filled the earth before the Flood. Its discourse of pleasure exculpates violence and cruelty in its strategy to remove the taint of sin from meat. This latter, says Vantassel (2009, 175), has the advantage of simplifying the moral ambiguity of killing animals, "as it removes the issue of animal suffering from the debate".

Meat and Masculinity

As we saw in previous chapters, hunting animals for pleasure produces and reproduces a discourse of masculinity; and violence towards animals prepares the warrior for warfare. The proof-texts contending that God likes "the smell of Barbecue" suggests a similar interaction between the pleasures of carnism and hegemonic masculinity. Indeed, hunters within evangelicalism use the language of carno-theology to establish their entitlement to kill, just as carnists use it to authorise their entitlement to eat flesh. In both cases human dominion is considered absolute, authorised by God. Carol Adams (2003) has drawn attention to the parallel between the sense of entitlement of men over animal bodies and misogyny, detailing many similarities between a culture of violence towards women and the portrayal of meat. Indeed, there is a striking similarity between the subordination of animal bodies to carnist entitlement, and the "subordination of women" to men allegedly authorised by God in the discourse of "complementarianism". The term "complementarianism" refers to the normative relationship between men and women within evangelicalism, asserting that it is hierarchically complementary. Just as carno-theology sidelines the impact of the Fall (Gen. 3), so also does complementarianism. This similarity extends from the use of proof-texts as justification, to specific features such as Adam's patriarchal authority to name both animals and Eve, supposedly demonstrating the subordination of both (Giles 2018, 61, 66, 76).

Carno-theology's sidelining of sin also endorses contemporary consumerism. From Edens' first act of over-consumption in eating forbidden fruit (Gen. 3), to the demand for meat rather than manna in the wilderness (Num. 11.4ff), nonconformist discourse condemned excess pleasure. These moments in the biblical narrative are largely absent from carno-theology, and with them restraint on excess. Carno-theology makes the animal defenceless before the hunger of man, whose lust to consume knows no limit. Like death, lust "consumes but is never the object of consumption". Carno-theology inaugurates a "sphere of absolute eating, of absolutely carnivorous virility"; it has become a theology of death (Clark 2016, 100).

Christian and Secular Dimensions

Carno-theology is a remarkably prominent aspect of evangelicalism. I have rarely met anyone identifying with that community who is unable to immediately quote several proof-texts against vegetarianism. Yet, the same person may be unable to cite even one for, say, the Trinity, a defining doctrine of Christianity. The latter, of course, is embedded in Church confessions, liturgy, and culture, in a way that carnism is not; in that sense, it requires no *proof*-text, although it has been *proved* in the biblical texts for nearly 2000 years. The prominence of carno-theology in evangelicalism suggests that vegetarianism continues to present a moral challenge to evangelicals in a way that the Trinity does not, requiring a knowledge of proof-texts for rebuttal.

I have focussed upon carno-theology in evangelicalism, but it has also become part of a wider secular debate deploring the human exploitation of creation generally, and animals in particular. In secular discourse, carno-theology is deployed to illustrate the allegedly baleful effects of religious superstition on the interests of animals, showing that "irrational" religious beliefs naturally produce irrational justifications for unethical behaviour. As Andy Hamilton (2004) ironically remarked, "The Bible clearly states that God made animals as ingredients". The disadvantage to this secular approach is that to accept the manifest claims of carno-theology, defensively enhances its status within evangelicalism. A more critical response to carno-theology would involve unveiling its latent appropriation of nonconformist discourse, tailored to justify the excess pleasures of carnism in the face vegetarianism's ethical challenge.

A Point of Resistance

Nothing establishes a boundary between humans and all other animals like eating them. Nonconformist discourse subverts that boundary, uniting the song of all living creatures in the praise and manifestation of God. Carno-theology authorises the boundary; when animals praise or manifest God in the bible, they become metaphor. Real animals are denied a relationship of praise to God in order to become meat. Sin is denied. The capacity for a nonconformist discourse to resist carno-theology is inadvertently recognised by Stephen Vantassel (2009, 156). If the purpose of animals is just "communing with God", what use are they to humans, he asks; "what 'work' [for humans] do animals perform?".

The nonconformist answer is "none"; the value of God's creatures does not depend upon their usefulness to sinful men. As I have argued above, each animal is of individual value in the eyes of God. Collective value resides in praise.

Carnism has become an ideology which colonises all aspects of life. This includes religion where carno-theology has become the language of choice in the influential constituency of evangelicalism; it also features prominently in the dominant discourse of secularism. The currently subjugated discourse of nonconformity retains resources to resist this, and can provide a more robust response.

At present, critical responses to carnism actually reinforce evangelicalism's carno-theology by accepting its manifest claims. When Henry Salt (1914, 31–2) wrote his pamphlet for the Vegetarian Society in 1914, he rejected the manifest claims of the carno-theology of his day. For Salt, those claims were a veil cloaking "the rank old Roman *Saturnalia* under the garb of Christian sentiment". Contemporary carno-theology is also a veil, concealing the face of carnist desire. Lifting the veil, we find a violent upending of the very nonconformist heritage claimed by evangelicalism, especially its vaunted adherence to the biblical text. The fact that appropriations of dominion have so often been co-opted by carnism should not prevent the identification and exposure of the strategies used. As discussed above, one strategy has been to transform the real animals of the bible into metaphors. This allows evangelicalism's adherents to proclaims that they join with all creatures to sing God's praise, as though they meant it. But outside the church walls, this proclamation is reduced to a figure of speech, allowing unfettered enjoyment of the excess pleasures of carnism. The disparity between the proclamation of faith, and the beliefs revealed by behaviour is clear. Yet, the secular acquiescence in the manifest claims of carno-theology has prevented challenge.

For nonconformists, creation's praise was not simply an uplifting thought to accompany singing in church, but a practical demand upon how they lived a life of piety. If there is any virtue in trying to interact with evangelicalism without reinforcing it, the nonconformist song of life calls us to witness against its violence, and to celebrate the lives of animals in the way we speak of them. It announces our human calling to conduct animals in their song of life, not consume them in a carno-theology of death.

The present chapter completes my discussion of the three areas most associated with animal cruelty and extreme suffering: hunting, warfare, and the meat industry. The next and final chapter provides a brief synopsis in the context of what the philosopher Jürgen Habermas (1992, 15, 51; 2002, 149) calls the "Judeo-Christian understanding of history in terms of salvation".

References

Adams, Carol J. 2003. *The Pornography of Meat*. New York: Continuum.

Adams, Carol. 2011. *The Sexual Politics of Meat* New York: Continuum.

Adams, Carol. n.d. What is or Could be the Christian Witness in the Face of Violence and Suffering? Sarx. https://sarx.org.uk/articles/food/witnessing-against-violence/. Accessed 29 April 2025.

Adam, Margaret B., David L. Clough, and David Grumett. 2019. A Christian Case for Farmed Animal Welfare. *Animals* 9: 1116.

Ainsworth, Henry. 1843 [1627]. *Annotations on the Pentateuch or the five books of Moses…*, vol. 1. Glasgow: Blackie & Son.

Anon. 2019. The Part Where God Says Its OK To Eat Bacon. Steppes of Faith, 26.7.2019. https://www.steppesoffaith.com/apologetics/god-says-okay-eat-bacon. Accessed 10 November 2024.

Anon. 2025a. Will there Be Animals In Heaven? Billy Graham Evangelistic Association, UK. https://billygraham.org.uk/answer/will-there-be-animals-in-heaven/. Accessed 29 April 2025.

Anon. 2025b. "Vegans, Vegetarians, and the Bible".. Focus on the Family. https://www.focusonthefamily.com/family-qa/vegans-vegetarians-and-the-bible/. Accessed 25 April 2025.

Anon. 2025c. Do Christians Have to Eat Meat? Jesus Everyday https://www.jesus-everyday.com/do-christians-have-to-eat-meat/. Accessed 25 April 2025.

Anon. 2025d. Vegetarianism http://www.blessedquietness.com/journal/housechu/vegechur.htm. Accessed 29 April 2025.

Antrobus, Derek. 1997. *A Guiltless Feast*. City of Salford Education and Leisure Department.

Bandura, Albert, Claudio Barbaranelli, Gian Vittorio Caprara, and Concetta Pastorelli. 1996. Mechanisms of Moral Disengagement in the Exercise of Moral Agency. *Journal of Personality and Social Psychology* 71 (2): 364–374.

Baxter, Richard. 1656. *Plain Scripture-proof of Infants Church-membership and Baptism*. London: T.V.F.T

Bayly, David and Tim. 2011. Feminism, Homosexism [sic], and Veganism: The Grand Conspiracy. December 27, 2011 Baylyblog http://baylyblog.com/blog/2011/12/feminism-homosexism-and-veganism-grand-conspiracy. Accessed 29 April 2025.

Bebbington, D. W. 1989. *Evangelicalism in Modern Britain: A History from the 1730s to the 1980s*. London: Unwin Hyman.

Bouma, Jeremy. 2015. *All God's Promises: The Ultimate Handbook for Every Need*. Wisconsin: Broadstreet.

Burgess, Anthony. 1658. *A Treatise of Original Sin... Proving that It Is, by Pregnant Texts of Scripture Vindicated from False* Glosses. London: s.n.

Calvin, John. 1845 [1559]. *The Institutes of the Christian Religion*. Trans. Henry Beveridge. Edinburgh: Calvin Translation Society.

Calvin, John. 1958. *Calvin's Commentaries*, vol. 23, trans. Joseph Haroutunian. Philadelphia: Westminster Press.

Calvin, John. 1965 [1578]. *Genesis*. London: Banner of Truth.

Calvin, John. 2018. *Harmony of the Law*, vol. 2, Lev. 3.16f. Woodstock, Canada: Devoted Publishing.

Clark, D. L. 2016. Hegel, Eating: Schelling and the Carnivorous Virility of Philosophy. In *Critical Perspectives on Veganism*, ed. J. Castricano, and R. R. Simonsen. Palgrave Macmillan, quoting John Caputo.

Cole, Matthew. 2008. Asceticism and Hedonism in Research Discourses of Veg*Anism. *British Food Journal* 110 (7): 706–716.

Croney, Candace, and Janice Swanson. 2023. Is Meat Eating Morally Defensible? *Animal Frontiers* 13 (2): 61–67.

Derrida, J. 1991. "Eating Well": An Interview with Jacques Derrida. In *Who Comes after the Subject*, ed. E. Cadava, P. Connor, and J.-L. Nancy, 96–119. New York: Routledge.

Derrida, J., and M. L. Mallet. 2008. *The Animal that Therefore I Am*. New York: Fordham University Press.

Ditzel 2011. Vegetarianism: Doctrine of Demons. Christian Forums. http://www.christianforums.com/t5986883. Accessed 21 March 2015.

Dunayer, Joan. 2001. *Animal Equality: Language and Liberation*. Ryce Publications.

Fee, Gordon D. and Douglas Stuart. 1993. *How to Read the Bible for All Its Worth*. Bletchley: Scripture Union.

Feucht, Sean. 2023. Facebook Post 21.9.2023. https://www.facebook.com/sean.feucht/posts/there-are-274-bible-verses-that-mention-meatthere-are-13-bible-verses-that-memti/866410828178791/. Accessed 22 November 2025.

Giles, Kevin. 2018. Complementarian Theology in Crisis. In *Eyes to See and Ears to Hear Women*, ed. Tim Krueger, 59–82. Minneapolis: CBE International.

Goldsmith, Oliver. 1870. *The Works of Oliver Goldsmith*. Edinburgh: William P. Nimmo.

Goodman, Michael K. and Colin Sage, eds. 2014. *Food Transgressions: Making Sense of Contemporary Food Politics*. Routledge.

Goodwin, Thomas. 1861. *Works of Thomas Goodwin*, vol. 4. Edinburgh: James Nichol.

Graça, João., Maria Manuela Calheiros, and Abílio. Oliveira. 2016. Situating Moral Disengagement: Motivated Reasoning in Meat Consumption and Substitution. *Personality and Individual Differences.* 90: 353–364.

Gummer, John Selwyn. 1990. Speech at The International Meat Trade Association dinner, held in London on May 1st 1990.

Habermas, Jürgen. 1992. *Post-metaphysical Thinking*. Cambridge: Polity.

Habermas, Jürgen. 2002. *Religion and Rationality*. Cambridge: Polity.

Hamilton, Andy. 2004. *Revolting People,* BBC Radio 4, June 2004.

Hodges, Thomas. 1675. *The Creatures Goodness….* London: Tho. Parkhurst.

Johnson, Lisa. 2017. On the Suffering of Animals in Nature. *Journal of Animal Ethics* 7 (1): 63–77.

Joy, Melanie. 2009. *Why We Love Dogs, Eat Pigs, and Wear Cows* Conari Press.

Joy, Melanie and Jens Tuider. 2016. The Rise of Veganism. In *Critical Perspectives on Veganism*, ed. Jodey Castricano & Rasmus Simonsen. Palgrave Macmillan.

Lane, Belden C. 2011. *Ravished by Beauty: The Surprising Legacy of Reformed Spirituality*. Oxford: OUP.

Liles, Jordan. 2025. Factcheck 10.9.2025. *Snopes*. https://www.snopes.com/fact-check/charlie-kirk-empathy-quote/. Accessed 25 September 2025.

Lloyd-Jones, D. M. 1992. *What is an Evangelical?* Edinburgh: Banner of Truth.

Luther, Martin. nd: 1525. *On the Bondage of the Will.* Grand Rapids: APA, S. 4.

Makin, Keith. 2024. *Independent Learning Lessons Review John Smyth QC.* 18th October 2024, Church of England.

Matherson, Dave. 2015. Eat Your Veggies… The Animal Rights Movement. http://www.tkc.com/resources/resources-pages/animalrights2.html, The Kings Community Church. Accessed 21 March 2015.

Midegley, Mary. 1998. *Animals and Why they Matter*. Athes, Georgia: University of Georgia Press.

Million Voices. 2024. God's Word: Handbook for Life. 27.11.2024 https://millionvoices.org/gods-word-handbook-for-life/. Accessed 26 April 2025.

Muller, Richard A. 2003. *Post-Reformation Reformed Dogmatics Vol 2: Holy Scripture*. Grand Rapids: Baker Academic.

Newell, Marcello. 2024. 'Faith Working Through Love': A New Food Law for a New World. *Theology* 127 (1): 22–29.

Otter, Chris. 2020. *Diet for a Large Planet*. University of Chicago Press.

Piazza, Jared, Matthew B. Ruby, and Steve Loughnan et al. 2015. Rationalizing Meat Consumption. The 4Ns. *Appetite* 91 (1): 114–128.

Pinker, Steven. 2012. *Better Angels of Our Nature*. Penguin.

Palin, Sarah. 2009. *Going Rogue*. HarperCollins.

Pollan, Michael. 2002. An Animal's Place. *New York Times* 10 November.

Rigney, Joe. 2025. *The Sin of Empathy*. Idaho: Canon Press.

Roberts, Mark. 2010. God's Favourite Smell. The High Calling 9.11.2010. http://www.thehighcalling.org/articles/daily-reflection/gods-favorite-smell. Accessed 5 February 2019.

Rothgerber, Hank. 2020. Meat-related Cognitive Dissonance: A Conceptual Framework for Understanding How Meat Eaters Reduce Negative Arousal from Eating Animals. *Appetite* 146: 104511.

Salt, Henry S. 1914. *The Humanities of Diet: Some Reasonings and Rhymings.* The Vegetarian Society Manchester.

Spurgeon, C. H. 1887. "Preface" *The Sword and the Trowel*, August 1887.

Stone, Ken. 2018. *Reading the Hebew Bible with Animal Studies.* Stanford University Press.

Stuart, Tristram. 2006. *The Bloodless Revolution.* Harper.

Swanson, Dennis M. 2001. The Down Grade Controversy and Evangelical Boundaries. *Evangelical Theological Society*, November 2001 at Colorado Springs, Colorado.

Thomas, Keith. 1983. *Man and the Natural World.* London: Allen Lane.

Vanhoozer, Kevin J. 2005. *The Drama of Doctrine.* Westminster: John Knox Press.

Vantassel, Stephen M. 2009. *Dominion Over Wildlife?* Oregon: Resource publications.

Wesley, John. 1931 [1747]. Letter to Dr. Gibson, Bishop of London, June 11, 1747. In *The Letters of John Wesley*, ed. John Telford, vol. 2. London: Epworth Press.

Wesley, John. 1872 [1771]. The General Deliverance. In *Sermons on Several Occasions*, vol. 2, Sermon 60.

Wilkinson, Loren. 1990. A Theology of Beasts. *Christianity Today* 34:9 (June 18, 1990), 21.

Žižek, Slavoj. 1989. *The Sublime Object of Ideology.* Verso.

Žižek, Slavoj. 2009. Notes on a Poetic-Military Complex. *Third Text* 23 (5): 503–509.

CHAPTER 11

Conclusion

Introduction

This book has traced the arc of an animal friendly way of speaking from its origin in the Puritans and Dissenters of the seventeenth century, to its subjugation in the new anthropocentric Evangelicalism of the later nineteenth century. Where the former had a passion for kindness, the latter came to promote hunting and carno-theology. I have suggested that, by recovering nonconformist discourse, we can progressively recenter our perception of modern animal ethics around the voice of manifestation and praise. The practical outworking of such singing together, argues Jürgen Moltmann (1995, 71), is "well suited to overcome the impoverished and one-sided attitudes of people living in the modern industrial world".

Of course, this nonconformist discourse was always in the minority, and its resistance to the dominant narrative was uneven. Yet it had an influence far beyond its numbers, and spoke *for* animals when few others did so. From the late sixteenth century, nonconformists developed a novel way of talking about animals which came to overwhelmingly dominate animal welfare reform (Preece, 2003, 401). In the eighteenth century, it was nonconformist Christians rather than Enlightenment philosophers or Romantic poets, who led the way. For example, David Perkins (2003, 44) considers that "no writer in the eighteenth century had more effect… in transforming attitudes to animals and stimulating reform" than the

P. J. Sampson, *Nonconformist Perspectives on Animals and Language*, The Palgrave Macmillan Animal Ethics Series,
https://doi.org/10.1007/978-3-032-12158-5_11

Calvinist poet William Cowper. Whereas the Romantic poets or Enlightenment philosophers were rarely cited in public debate, Cowper was "quoted over and over in sermons, pamphlets, and in Parliament".[1] We might add Cowper's popularity among ordinary Christians as they sang his hymns each Sunday; and among the more literary who read him, posthumously, in the work of Jane Austen or Anne Brontë(Brooke 1999, 127–8; Brontë 1846, 92–4; Sampson, forthcoming). This continued to enrich ways of speaking *for* animals throughout the eighteenth century.

The nineteenth century saw a similar pattern. The Victorian animal welfare Acts, and organisations to prevent cruelty such as the RSPCA, had their origin in the nonconformist evangelical religion of the time, drawing inspiration from the biblical texts (Preece 2005, 28). Preece and Chamberlain (1993, 38; also 40) remark that "[i]ronically, it was those who repudiated Darwin's elevation of the status of animals who, in practice, did most to protect the interests of animals". They did this by using a language of wonder and joy. I discussed several examples of this in the latter part of this book. Some debates, such as the ethics of hunting and diet, are now well established public discourses; others, such as the exploitation of animals in warfare, remain largely unexplored.

Animal Advocacy and Religion

Of course, nonconformist discourse is not the only "religious" language capable of sustaining animal friendly speech (Linzey and Linzey 2019). However, it has been the most influential in the Western ethical tradition. Before its subjugation, it contributed uniquely to animal friendly ways of talking for over two centuries. Serious as its decline was, it did not amount to a total eclipse. For example, during the evangelical Welsh revival of 1904, it was said that a converted miner could always be recognised because his pit pony did not flinch when he raised his hand (Matthews 2004, 55). Pit ponies were no longer property, dispensable beasts of burden; they had become creatures of God.

It is widely recognised that the dominant discourse of animals as property has been far less successful at developing an animal friendly way of speaking. However, simply turning from materialist to a "religious" language is not necessarily a panacea. As Keith Thomas (1984, 17f,

[1] For Cowper's popularity, see Elfenbein (1999, 63, 158).

25) has documented, the Tudor and Stuart age commonly exhibited an "aggressively despotic" language of "mastery" and "dominion", claiming justification in the biblical texts, especially the first chapter of Genesis. Nonconformist language was often appropriated and transformed into something alien to the original. This has certainly continued into the modern era. We have already noted that the sometime member of The Church of England General Synod, John Selwyn Gummer, considered that the "Bible tells us that we are masters" over the animals, and so "we very properly eat them". This appropriation of religious language by the dominant discourse is not restricted to Christianity.

"Indigenous teachings" about "the interconnection of all of life" in traditional societies have been cited as providing sources of wisdom (Saito 2023, 4; Kimmerer 2013). Indeed, Sidney Blankenship (2019) has argued that just such wisdom provides a resource for "restoring species to the Circle of Life" following the "catastrophic disruption" of ecosystems in North America. This authentic wisdom is sometimes appropriated by the subsidiary Western discourses associated with Rousseau and the Romantic movement (Kimmerer 2013, 6f). Such appropriation has entailed a degree of generalisation and idealisation which are no more reliable than those about Western Christianity (Preece 1999, 94f 214f, and ch 7). For example, Pinker (2012, 551-2) records a Native American recipe for roast turtle which has the simple instructions: "Put a turtle on his back on the fire. When you hear the shell crack, he's done". I argued in Chapter 9 that specifically *religious* language can be especially effective in justifying activities which would otherwise appear wholly reprehensible. This applies to all such language, not just that of anthropocentric Evangelicalism.

"Spiritual" but less formally "religious" language need not be much better. Many "non-religious" people recognise the harmony and unity of nature, and often describe this as *spiritual*. For example, Ernest Hemingway (1923, 197, 234) possessed a keen sense of "spiritual intensity" in the unity between humans and other animals, writing of the "emotional and spiritual intensity, and pure, classic beauty that can be produced by a man, an animal, and a piece of scarlet serge draped over a stick". But he was, of course, referring to the bullfight, where "the beauty of the moment of killing is that flash when man and bull form one figure as the sword goes all the way in, the man leaning after it, death uniting the two figures in the emotional, aesthetic, and artistic climax of the fight".

Harmony, community, and spiritual intensity are not enough to generate an animal friendly language; indeed, they may do quite the opposite. As I argued in Chapter 5, the nonconformist discourse of singing together recognises such risks, and provides explicit safeguards against cruelty.

A Relish of Salvation

I have argued elsewhere that traces and vestiges of nonconformist discourse remain in contemporary debates, and continue to show a capacity to speak *for* animals in the Western social imaginary (Sampson 2018, ch. 8). Indeed, the contemporary philosopher Jürgen Habermas (1992, 15, 51; 2002, 149) has argued that the Western tradition has no alternative heritage to that of the "Judeo-Christian understanding of history in terms of salvation", and that it is "inspiring and even indispensable". This is not least because, as a substantially unchanged legacy which has been the "object of continual critical reappropriation and reinterpretation" for two thousand years, it is unique, at least in the Western intellectual tradition.

Habermas' emphasis on "salvation" is significant for our discussion. Whenever animal advocates appeal to tropes of protection or safeguarding, they invoke a relic of salvation. Even as secular animal advocacy holds "dominion" responsible for the anthropocentrism which it condemns, it unconsciously draws on the biblical vision which it has already appropriated. Calls to "Save the Whales" imply that whales matter, that it is both desirable and possible to save them. This tacitly assumes the communal song of praise in which all creatures matter, not just in the opinion of the speaker, but as a transcendent demand on everyone. There is, as Shakespeare (Hamlet Act 5.2.233) put it, a "special providence in the fall of a sparrow"; God cares for them. Sparrows matter, even though humans care little for them, and "sell two for a farthing"; how much more does "everything that he [God] had made" matter (Matt. 10.29; Gen. 1.31).

The relics of salvation which persist in our culture retain cultural force, and are already part of the cultural landscape requiring no further justification. They carry meanings which inspired a striking capacity over several hundred years to find value in animal life. Yet they need nurturing and refreshing if their ethical force is to persist. While "many figures of our own culture are... thoroughly derived from biblical patterns and rhythms", their origin is "largely unknown, misunderstood or condemned", argues Christopher Watkin (2022, 14-15). In the case of animal ethics, this leaves them open to appropriation by less animal friendly discourses. We cannot assume that the underlying biblical "patterns and rhythms" will be self-sustaining. Once the grand narrative of salvation fades from view, we are left with only fragmented communities of opinion.

Moreover, other narratives exist which might call the relish of salvation into question. Some human culinary traditions and economic activities have shown themselves both widely influential and utterly indifferent to saving whales, sparrows, or any other exploitable animal. Within a Darwinian universe of survival of the fittest, it might be that whales will, and should, go the way of the dinosaur.

Nonconformist discourse provides resources for the animal advocacy community to articulate meanings which exceed pragmatic reason and positivist explanations. Moreover, as Habermas argues, the "Judeo-Christian understanding of history in terms of salvation" also provides a source of ethical inspiration. New issues affecting animals constantly arise, and those most in need of urgent discussion are those for which a language doesn't quite exist. Animals in warfare, discussed in Chapter 7, is a case in point. Nonconformist discourse provides a source of fresh insights to enrich the imagination of both popular and technical animal advocacy. In particular, it can provide distinctive resources for the critical review of those traditions which are indifferent to salvation. Indeed, if Habermas is correct, it is "indispensable". There are really no other option.

Today, nonconformist vestiges remain part of contemporary public discourse. But no-one should take the "relish of salvation" for granted (Hamlet Act 3.3.97). It would be wise to "strengthen the things which remain" (Rev. 3.2). If we lose a way of saying something, we are no longer able to say it. This book has been a modest attempt to remind us of a rich but little appreciated heritage of speaking *for* animals; of *caring* for animals, even sparrows.

The Witness of Care

Throughout this book, the theme of care has frequently recurred as tacit in the song of creation, and central to the art of conducting. It remains important. For example, many contemporary Christians use a language of "creation care" or "stewardship", though usually in relation to environmentalism rather than animals (Berry 2000).

Jesus of Nazareth more often spoke of care for animals than for the environment, not least because he lived long before the anthropocene in which humans have become the global force for environmental change. For example, he cared enough to keep the ass and her colt together as he rode into Jerusalem; and he rebuked those who cared little for sparrows; a reproach which has lost its force as it has gained in familiarity (Matt. 21.2, 10.29; Luke 12.6). No animal is "too mean" for care, said the puritan John Maynard (1688, 19-20): "If I should name some [cruel] practices among us..., perhaps some would think them too mean to be mentioned here.... [The] poorest Creature that is, is a part of Gods Workmanship, and God may be dishonoured in it: yea, to abuse in sport (though it be as mean a Creature as a Flie &c.) is to take the Name of God in vain; since these works of *His,* are such things, by which, as by a name, he hath made himself known unto us; the least Flie being an evidence of his Almighty Power".

Cowper (1849:1798, vol. 2, 138; book VI, lines 560-7) famously declined to befriend someone who "needlessly sets foot upon a worm"; the theologian and humanitarian, Albert Schweitzer, saved an insect from a puddle (Jorgensen 2025, section 7); Spurgeon (1885, 148.10) and Manton (1870, vol. 1, 443) echoed Maynard's wonder at a fly. Such micro-caring does not necessarily indicate an over-sensitivity to suffering, or even a tacit assertion that worms and insects can suffer at all; but rather a recognition of the glory of God and a witness of care to others. I have a friend who stopped a group of boys from "tormenting" an insect, not because he thought the insect suffered, but because the boys thought so. Witnessing is performative, not in the theatrical sense, but in that it makes a difference in the world (Austin's 1975). It calls others to care. Anders Jorgensen (2025, section 7) speaks of the witness of care, inspiring others to also act caringly. A caring act "becomes a symbol of the possibility of caring relations, ... of kindness or justice, and as such a carrier of hope".

Care is not an abstract or complex criterion for compassion. If we care, we watch and listen; everyone who does so can tell the difference between a lamb gambolling in the field and a lamb screaming under the butchers knife. As discussed in Chapter 5, this is most obvious with mammals whose cries we readily understand. Indeed, Alexander Pope (1713) suggested a deliberate divine strategy: "[p]erhaps that voice or cry [which] so nearly resembles the human, with which Providence has endued so many different animals, might purposely be given them to move our pity, and prevent those cruelties we are too apt to inflict on our fellow creatures". The early naturalist, Thomas Bewick (1862, 8), tells that as a boy he abandoned the cruelty of hunting when he caught a hare in his arms "while surrounded by the Dogs & the Hunters, when the poor terrified creature screamed out so piteously, like a child, that I would have given any thing to save its life". Not all animals speak so eloquently. Some, like fish, have a voice which we are less well attuned to hear. Yet, as was discussed in Chapter 2, in the nonconformist vision, fish also "are vocal" with praise, "they have a speech though they have no voice". Caring implies value; it is a witness that the world carries meaning, and a "carrier of hope".

Of course, many close their ears against an animal's cries, turn away from empathy. For some, this is simple hard-heartedness, as discussed in Chapter 5. For example, Bewick gave his hare to a nearby farmer to save her life, but the farmer desired "to have a bit more sport with her", which was done "by first breaking one of her legs, and then again setting the poor animal off a little before the dogs". Remarkably, the hare escaped. Sometimes, hard-heartedness is not enough, and we need to make a special effort to silence empathy and care. In Chapter 9, we saw the strategies adopted by carno-theology to do just that. It seems that ordinary rationalisations are not enough to authorise the greatest cruelties; the extraordinary claims of religion are required.

Salvation and Hope

Empathy and care can be painful; perhaps this is the reason that so many harden their hearts against compassion for animals, or regard empathy as a sin. Simply preserving dappled patches of compassion for the animals closest to us is hard enough; consistently advocating for all animals requires a special commitment. Speaking for animals comes at a cost and requires grace says Linzey (1998, 7-8): "I thank God for the grace to feel their suffering and give voice to their pain… May God the Holy Trinity give me strength to live out my commitment this day". This was no less the case for the nonconformist pioneers of reform who generally lacked the support now available to animal advocates. Yet they persevered in the face of opposition and ridicule.

Animal advocates may have complex motivations including health, the environment, and compassion; but most share the central conviction of a moral universe, and a vision of the future which sustains their hope for the long haul in the face of adversity. "The arc of the moral universe is long, but it bends toward justice" quoted Martin Luther King, inspiring hope in the civil rights movement of the mid-twentieth century. Animal advocates build their hope upon a moral universe which bends towards justice, love, and compassion. Of course, this can be appropriated by the dominant discourse in order to sideline animal advocacy. As Peter Singer (1975, 215, ix) has pointed out, "portrayal of those who protest against cruelty to animals as sentimental, emotional "animal-lovers" has had the effect of excluding the entire issue of our treatment of nonhumans from serious political and moral discussion". He here hints at the kind of gendered language used to oppose compassion which was discussed in Chapter 5. Nonconformists shared with modern animal advocates the hope for a future which is greater than the present. They articulated their moral vision in a language representing *restoration,* not simply preservation or maintaining the status quo. In fact, they looked to a future which was greater than the past, "much more beautiful and joyful than paradise was", as Luther says (Santmire 1985, 131). Its occasional gendered appropriation rarely deterred them.

Restoration inspired hope for substantial change: there is, said the Scottish presbyterian, David Dickson (1659, 21), a "hope of the Creatures, that they may bee freed from that servitude into a state proportionable to the future glorious condition of the Sons of God". As the puritan, Walker (1641, 168) explains, in the state of innocency "all creatures lived together in love and concord among themselves…" "And blessed shall the times bee, when God shall restore the creatures againe from this bondage and corruption to the glorious liberty of the sons of god, as hee hath foretold by the holy prophet Is. 11.6". The reference to Isaiah concerns the prophecy that, in the restored peaceable kingdom, the wolf "shall dwell with the lamb" (see also Is. 65.25, Hos. 2.18, Acts 3.21 etc.).

The puritan, Charnock (c1660:1864, 348), reminded his readers that the "last time is called not a time of destruction, but a 'time of restitution', and that of all things, Acts iii.21", when the "fierceness" of creatures among themselves will vanish. "The world shall be nothing but a universal smile. Nature shall put on triumphant vestments".

As discussed in Chapter 1, this moral vision was understood as not so much mere metaphor, but as a substantial reality bending towards peace. It provided hope for an end to violence, recognising and valuing a common vulnerability, care, and joy. It affirmed the hope that their "labour is not in vain" (1 Cor 15.58). The world is "transformable in the direction of its promised future, and so it rouses active hope which seeks out possibilities of change and creates anticipations of the future kingdom of God" (Bauckham 1993, 386). It motivates and sustains commitment to singing together even while we share the pain of dissonance, building symbols for the future.

By making the peaceable kingdom a feature of a substantial reality, this moral vision has shown a remarkable capacity to express meanings beyond the pragmatic use of animals; an ability to find a bond of solidarity between animals and humans without effacing difference (Levinas 1969, 40); and a soteriologically rooted hope that we humans will treat animals less cruelly in a world, if not restored, at least without those who "fill [it] with riot and defile [it] with [the] blood" of animals. Such a peaceable "[p]aradise (for earth has still, some traces of her youthful beauty left)" anticipates "the times of restitution of all things, which God hath spoken by the mouth of all his holy prophets since the world began", when the "wolf also shall dwell with the lamb" (Cowper 1849:1798, book 3. 307f, 298–99; vol.2, 60). This hope has not been idle, but has historically inspired social and political action to make the world less exploitative (Berry 2015, 180f, Sampson 2018, 47f).

Without such a vision of restitution, animal advocates tend to advocate for disengagement from free-living animals, leaving them to the mercilessness of Darwinian survival of the fittest. This was always problematic, and has become increasingly so as humans have extended their sway over nature in the anthropocene. I have argued elsewhere that "rewilding" projects which disengage in this way neglect the fear and suffering which the introduction of species may cause (Sampson 2021).

This vision remains in the mainstream of Christian worship. C. S. Lewis (2004:1958, 955-6) wrote a hymn for Easter, a time when the Christian church celebrates the risen Prince of heaven who has "given all who will the hope of heaven", including all animals.

Rise and dance, ye ransomed legions
Of the cold and countless dead!
Gates of adamant are broken,
Words of conquering power are spoken
Through the God who died and bled
Hell lies vacant, spoiled and cheated
By the Lord of life defeated.
Bear, behemoth, bustard, camel,

Warthog, wombat, kangaroo,
Insect, reptile, fish and mammal,
Tree, flower, grass, and lichen too,
Rise and romp and ramp, awaking,
For the age-old curse is breaking.
All things shall be made new.

This is no over-realised eschatology. Rather, in Bauckham's words, it defines the "direction of a promised future", rousing hope and inspiring a culture of advocacy. To live in a way which anticipates the restoration of all things is readily recognised by most animal advocates. If there is anything to the nonconformist prophetic motifs of wolves and lambs, it also puts them on the right side of history.

References

Austin, J. L. 1975. *How to do things with words*. OUP.

Bauckham, Richard. 1993. Moltmann, Jürgen. In *Modern Christian thought*, ed. Alister E. McGrath, 385–388. Oxford: Blackwell.

Berry, Evan. 2015. *Devoted to nature: The religious roots of American environmentalism*. San Francisco: University of California Press.

Berry, R. J. 2000. *The Care of Creation*. Leicester: IVP.

Bewick, Thomas. 1862. *A memoir of Thomas Bewick written by himself*. London: Longman et al.

Blankenship, Sidney. 2019. Native American religion. In *Linzey and Linzey*, 116–126.

Brontë, Anne. 1846. To Cowper. In *Poems*. London: Smith, Elder and Co.

Brooke, C. N. L. 1999. *Jane Austen: Illusion and reality*. Brewer: D. S.

Charnock, Stephen. c1660:1864. A Discourse on the existence and Attributes of God. In *The Complete Works of Stephen Charnock* Edinburgh: James Nichol, vol. 2.

Cowper, William. 1849:1798. The task.... In Poems. Edited by John Johnson. London: Phillips, Sampson & Co.

Dickson, David. 1659. *An exposition of all St. Pauls epistles....* London: Francis Eglesfield.

Elfenbein, Andrew. 1999. *Romantic genius: The prehistory of a homosexual role*. Columbia University Press.

Habermas, Jürgen. 1992. *Post-metaphysical thinking*. Cambridge: Polity.

Habermas, Jürgen. 2002. *Religion and rationality*. Cambridge: Polity.

Hemingway, Ernest. 1923. *Death in the afternoon*. London: Jonathan Cape.

Jorgensen, Anders Skou. 2025. Care in Times of crisis: Phenomenological, political and theological perspectives. *Dialogue* 64 (1): 36–44.

Kimmerer, Robin Wall. 2013. *Braiding Sweetgrass: Indigenous wisdom, scientific knowledge, and the teachings of plants*. Minneapolis MN: Milkwood.

Levinas, Emmanuel. 1969. *Totality and Infinity*. Pittsburgh: Duquesne University Press.

Lewis, C. S. 2004:1958. My Easter hymn. In The collected letters of C. S. Lewis. Edited by Walter Hooper. HarperCollins, vol 3.

Linzey, Andrew. 1998. *Animal Gospel*. Louisville, KY: Westminster John Knox Press.

Linzey, Andrew, and Clair Linzey. 2019. *The Routledge handbook of religion and animal ethics*. Routledge.

Manton, Thomas. 1870. *The complete works*. London: James Nisbet.

Matthews, David. 2004. *I saw the welsh revival*. Belfast: Ambassador.

Maynard, John. 1688. *The beauty and order of the creation*. London: William Gearing.

Moltmann, Jürgen. 1995. *God in creation*. Minneapolis: Fortress Press.

Perkins, David. 2003. *Romanticism and animal rights*. Cambridge University Press.

Pinker, Steven. 2012. *Better angels of our nature*. Penguin.

Pope, Alexander. 1713. *Against barbarity to animals*. Vol. 61. London: The Guardian. May 21.

Preece, Rod. 1999. *Animals and nature*. Vancouver: UBC Press.

Preece, Rod. 2003. Darwinism, Christianity, and the great vivisection debate. *Journal of the History of Ideas* 64 (3): 399–419.

Preece, Rod. 2005. Introduction. In *Immortal animal souls*, ed. Rod Preece. Lewiston NY: E. Mellen Press.

Preece, Rod, and Lorna Chamberlain. 1993. *Animal welfare and human values*. Ontario: Wilfrid Laurier Univ. Press.

Saito, Kohei. 2023. *Marx in the Anthropocene*. Cambridge: CUP.

Sampson, Philip. 2018. *Animal ethics and the nonconformist conscience*. Palgrave Macmillan.

Sampson, Philip. 2021. Noah, Jesus and ecology—sustaining and restoring the environment. In *Climate crisis and sustainable creaturely care*, ed. Christina Nellist, 137–151. Cambridge: CSP.

Sampson, Philip. Forthcoming. *A Passionate Kindness*.

Santmire, H. Paul. 1985. *The travail of nature*. Philadelphia: Fortress Press.

Singer, Peter. 1975. *Animal liberation*. New York Review Books.

Spurgeon, C. H. 1885. *The Treasury of David*. London: Marshall Brothers.

Thomas, Keith. 1984. *Man and the natural world*. Penguin.

Walker, George. 1641. *God made visible in his works*. London: John Bartlet.

Watkin, Christopher. 2022. *Biblical critical theory*. Grand Rapids Michigan: Zondervan.

Name Index

P. J. Sampson, *Nonconformist Perspectives on Animals and Language*, The Palgrave Macmillan Animal Ethics Series,
https://doi.org/10.1007/978-3-032-12158-5

Subject Index

P. J. Sampson, *Nonconformist Perspectives on Animals and Language*, The Palgrave Macmillan Animal Ethics Series,
https://doi.org/10.1007/978-3-032-12158-5

Kolibri 360 GmbH
Ferdinand-Jühlke-Straße 7
99095 Erfurt, Deutschland
produktsicherheit@kolibri360.de